Introduction to Air Pollution

Introduction to Air Pollution

Edward Campbell

STATES
ACADEMIC PRESS
www.statesacademicpress.com

Introduction to Air Pollution
Edward Campbell
ISBN: 978-1-63989-299-0 (Hardback)

STATES
ACADEMIC PRESS

Published by States Academic Press,
109 South 5th Street,
Brooklyn, NY 11249, USA

Cataloging-in-Publication Data

Introduction to air pollution / Edward Campbell.
p. cm.
Includes bibliographical references and index.
ISBN 978-1-63989-299-0
1 Air--Pollution. 2. Air quality. 3. Air quality management. 4. Atmospheric deposition.
I. Campbell, Edward.
TD883 .I58 2022
363.739 2--dc23

Table of Contents

Preface

Air pollution is defined as the addition of unwanted contaminants in the air which have adverse impact on human health. These contaminants are also referred to as pollutants. These pollutants are released by human activities as well as natural sources. Some prominent sources of pollutants are factory smoke, car emissions, pollens, volcanic eruptions, etc. Pollutants can be classified into primary pollutants and secondary pollutants. Primary pollutants include gases such as $CO2$ which is released due to indiscriminate burning of fossil fuels by humans and is responsible for the greenhouse effect. Other primary pollutants, namely, SOx and NOx are responsible for pulmonary disorders in humans and acid rain which inevitably leads to premature death. Secondary pollutants include ground level ozone ($O3$) and peroxyacetyl nitrate ($C2H3NO5$), both of which are formed by primary pollutants NOx and VOC's or volatile organic compounds. It has been estimated that pollution costs the world $5 trillion a year and results in around 7 million deaths annually, making it the single largest environmental risk. Some of the alternatives to reduce air pollution could be use of biofuels, electric motor vehicles and solar energy for generating electricity. Different approaches, evaluations and methodologies on air pollution have been included in this book. This book presents the complex subject of air pollution in the most comprehensible and easy to understand language. Those in search of information to further their knowledge will be greatly assisted by this book.

Given below is the chapter wise description of the book:

Chapter 1- The presence of a mixture of solid particles and gases in the air that are harmful to the health of living beings and cause climate change is known as air pollution. Air pollutants are airborne substances that cause air pollution and toxify the environment. This chapter serves as an introduction and guide to the vast subject of air pollution.

Chapter 2- The microscopic particles of solid and liquid matter present in the air are commonly known as atmospheric aerosol. The process in which a small number of ions, atoms or molecules becomes arranged in a pattern like that of a crystalline solid and form a site upon which further particles are deposited is known as nucleation phenomenon. This chapter discusses in detail atmospheric aerosols and their aspects.

Chapter 3- Air quality is the degree of how polluted or how clean the air is. Air Quality Index is used for reporting daily air quality. There is a system to monitor various parameters of air such as the concentration of air pollutants, wind speed and direction, etc. which is known as an air quality monitoring system. This chapter has been carefully written to provide an easy understanding of the varied facets of air quality.

Chapter 4- Various Air pollution mitigation techniques and devices such as Thermal Oxidizer, Catalytic oxidizer, Emissions Trading, Pollution Tax, and Dust Collection System are discussed in this chapter. It provides detailed insights on globally accepted policies and laws for air pollution mitigation.

Chapter 5- Highly reactive gas composed of three oxygen atoms is called ozone. It is of pale blue color and pungent smell. Ozone depletion can be termed as breaking of ozone layer which covers the topmost layer of earth's atmosphere. This chapter examines Ozone Cycle, Ozone Layer Depletion, CFCs, Ozone Hole and its consequences in a detailed manner.

Indeed, my job was extremely crucial and challenging as I had to ensure that every chapter is informative and structured in a student-friendly manner. I am thankful for the support provided by my family and colleagues during the completion of this book.

Edward Campbell

1

Air Pollution and Pollutants

The presence of a mixture of solid particles and gases in the air that are harmful to the health of living beings and cause climate change is known as air pollution. Air pollutants are airborne substances that cause air pollution and toxify the environment. This chapter serves as an introduction to the subject of air pollution.

Air pollution occurs, when harmful and toxic gases, dust and smoke, etc, enter the atmosphere. These make the survival of humans, plants and animals difficult and also disfigure the buildings and lead to deterioration of assets. According to Brimblecombe, air pollution occurs when the substances are released into the air by human activities in such concentrations as are sufficient to cause detrimental effect on human health, vegetation, animals and property or interfere with the enjoyment of the property by the mankind. So the contamination of air whether inside or outside the house is the cause of air pollution. These cause physical, biological or chemical changes in air, which if harmful, cause air pollution. The composition of air is given in Table.

Table: Composition of dry unpolluted air.

Gases	Per cent
Nitrogen	78.084%
Oxygen	20.946%
Argon	0.934$%
Carbon dioxide	0.039%
Neon	0.0018%
Helium	0.000526%
Methane	0.00016%
Krypton	0.00005%
Hydrogen	0.00003%
Nitrous oxide	0.00003%
Ozone	Trace to 0.0008%
Carbon monoxide	Trace to 0.000025
Sulfur dioxide	Trace to 0.00001%
Nitrogen dioxide	Trace to 0.000002%
Ammonia	Trace to 0.0000003%

It is remarkable to note that the concentration of pollutant gases is less than 0.1% but their toxic effects are highest. Out of the list in Table, the pollutant gases are tropospheric ozone, nitrogen oxides, sulfur dioxide, carbon monoxide, ammonia, methane, etc.

Causes of Air Pollution

- Burning of Fossil Fuels: The combustion of fossil fuels emits a large amount of sulphur dioxide. Carbon monoxide released by incomplete combustion of fossil fuels also results in air pollution.
- Automobiles: The gases emitted from vehicles such as jeeps, trucks, cars, buses, etc. pollute the environment. These are the major sources of greenhouse gases and also result in diseases among individuals.
- Agricultural Activities: Ammonia is one of the most hazardous gases emitted during agricultural activities. The insecticides, pesticides and fertilizers emit harmful chemicals in the atmosphere and contaminate it.
- Factories and Industries: Factories and industries are the main source of carbon monoxide, organic compounds, hydrocarbons and chemicals. These are released into the air, degrading its quality.
- Mining Activities: In the mining process, the minerals below the earth are extracted using large pieces of equipment. The dust and chemicals released during the process not only pollute the air, but also deteriorate the health of the workers and people living in the nearby areas.
- Domestic Sources: The household cleaning products and paints contain toxic chemicals that are released in the air. The smell from the newly painted walls is the smell of the chemicals present in the paints. It not only pollutes the air but also affects breathing

Effects of Air Pollution

The hazardous effects of air pollution on the environment include

- Diseases: Air pollution has resulted in several respiratory disorders and heart diseases among humans. The cases of lung cancer have increased in the last few decades. Children living near polluted areas are more prone to pneumonia and asthma. Many people die every year due to the direct or indirect effects of air pollution
- Global Warming: Due to the emission of greenhouse gases, there is an imbalance in the gaseous composition of the air. This has led to an increase in the temperature of the earth. This increase in earth's temperature is known as global warming. This has resulted in the melting of glaciers and an increase in sea levels. Many areas are submerged underwater.
- Acid Rain: The burning of fossil fuels releases harmful gases such as nitrogen oxides and sulphur oxides in the air. The water droplets combine with these pollutants, become acidic and fall as acid rain which damages human, animal and plant life.
- Ozone Layer Depletion: The release of chlorofluorocarbons halons, and hydro chlorofluor - carbons in the atmosphere is the major cause of depletion of the ozone layer. The depleting ozone layer does not prevent the harmful ultraviolet rays coming from the sun and causes skin diseases and eye problems among individuals.
- Effect on Animals: The air pollutants suspend on the water bodies and affect the aquatic life. Pollution also compels the animals to leave their habitat and shift to a new place. This renders them stray and has also led to the extinction of a large number of animal species.

Anthropogenic Air Pollution Sources

The anthropogenic chemical pollution has no borders and no matter where the pollutants are released into the atmosphere will have an impact over global environment. The most relevant sources are the incineration of fossil fuels to produce energy (heat and electricity), major industrial processes (like metallurgy industry or cement/construction industry) and transportation. We will classify the anthropogenic chemical pollution sources into two major groups: Stationary and mobile sources.

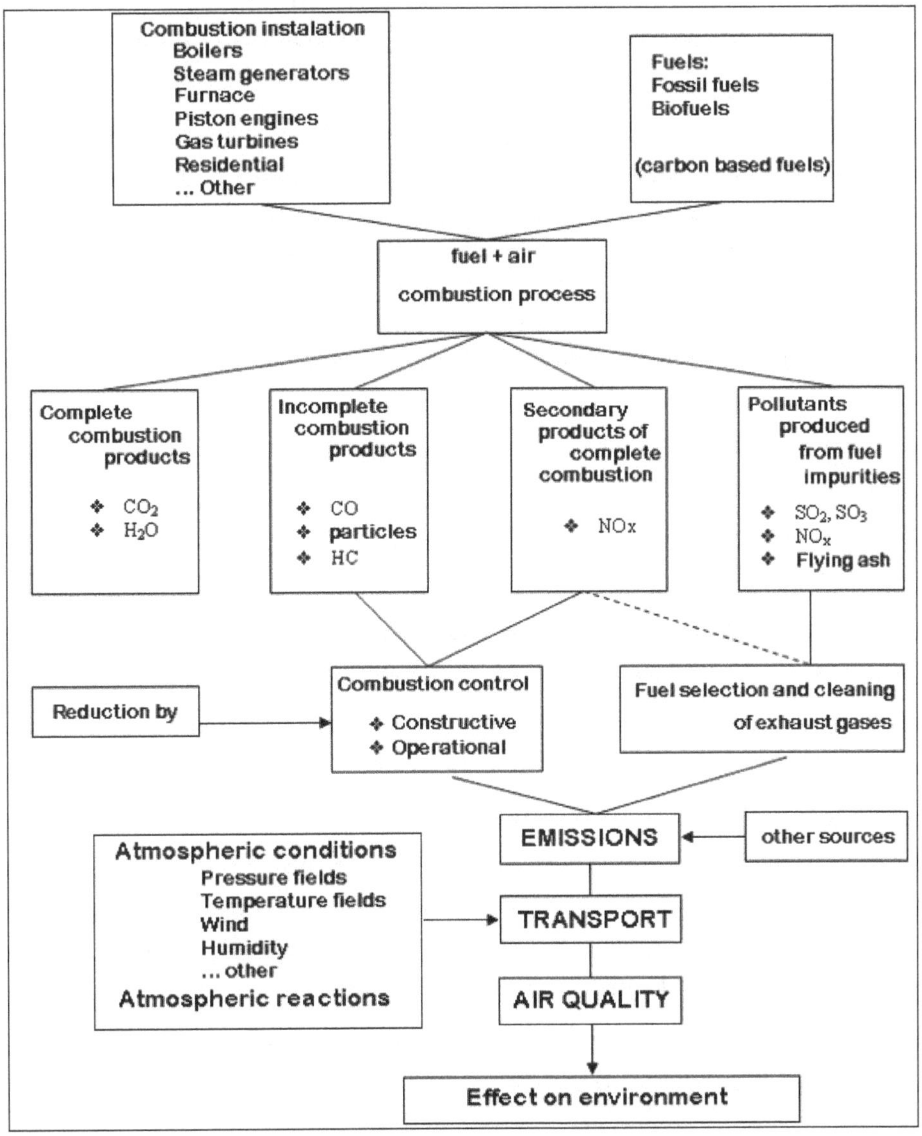

Sources and products of anthropogenic pollution.

The four major groups of gaseous air pollutants by historical importance, concentration, and overall effects on plants and animals (including people), are sulphur dioxide (SO_2), oxides of nitrogen (NOx: NO, NO_2), carbon dioxide (CO_2) and ozone (O_3). Sulphur dioxide and nitric oxide (NO) are primary pollutants – they are emitted directly from sources.

The most important groups of anthropogenic air pollution sources are defined by industrial processed, residential heating systems, transportation (terrestrial, naval and aerial) and agricultural systems. The majority of the pollutants are the direct result of the combustion process in large power plants and piston engines so that the first steps to reduce the pollutants concentration from atmosphere is to control and reduce the emissions from those source groups.

In the ideal case the combustion process is complete (perfect) and the exhaust gases are formed only from CO_2 and H_2O. To achieve the complete combustion is necessary to obtain a combustion air-fuel ratio (different for each type of combustion chamber) constant into the whole combustion chamber and this is not possible due to limitations is the construction of the combustion chamber. In Figure a short and simplified relation between fuel, combustion, combustion products (pollutants) and air quality is given.

The fossil fuels, solid, liquid or gaseous, are mainly formed by carbon and hydrogen in various ratios. During the combustion of this fuels the carbon and hydrogen reacts with oxygen to form carbon dioxide and water. Due to incomplete combustion or other substances included in fuels, like nitrogen, sulphur, organic compounds, heavy metals and other compounds, besides CO_2 and H_2O other chemical compounds will be formed, compounds that are dangerous both to humans and environment if their concentrations in ambient air exceeds supportability limits. Equation below shows the products of complete combustions of an ideal fuel formed only from carbon and hydrogen:

$$C_nH_m + \left(n + \frac{m}{4}\right)O_2 \rightarrow nCO_2 + \frac{m}{2}H_2O$$

From an ecological point of view we can distinguish between several types of air pollutants, depending on how they affect the ecosystem

- Pollutants that will have a direct impact on human health, like nitrogen oxides (nox), carbon monoxide (CO), sulphur oxides (SO_2) or volatile organic carbon (VOC).
- Pollutants with direct impact on vegetation, like nitrogen oxides (NO_x), sulphur oxides (SO_2) and chemical compounds of chlorine and hydrogen.
- Pollutants that will form acids, like (NO_x) and sulphur oxides (SO_2), especially when they are found in high concentrations and in same time into a humid atmosphere.
- Persistent pollutants with a long life cycle and accumulated in soil and by the transfer through biological plant-animal-human chain or accumulations in human body will have serious consequences on human health.
- Pollutants with direct influence over climate, like carbon dioxide (CO_2) and methane (CH_4) with major impact on global warming issues.

Carbon Monoxide (CO)

Carbon monoxide is an odorless, colorless and toxic gas. Because it is impossible to see, taste or smell the toxic fumes, CO can kill you before you are aware it is in your home. At lower levels of exposure, CO causes mild effects that are often mistaken for the flu These symptoms include headaches, dizziness, disorientation, nausea and fatigue. The effects of CO exposure can vary greatly from person to person depending on age, overall health and the concentration and length of exposure.

Carbon monoxide is mainly produced as an intermediary product of combustion processes in piston engines and boilers. The oxidation of CO into CO_2 requires a temperature of minimum 990 K and a sufficien stationary time into the combustion chamber. If the combustion temperature is not high enough or uniform inside the combustion chamber, when the exhaust gases are evacuated some CO will not oxidize. The dependency between some pollutants and oxygen content of fuel mixture and excess air ration is given in Figure.

Ambient CO concentrations in metropolitan areas are orders of magnitude higher than background. These levels are primarily associated with transportation emissions and are closely related to traffic density and meteorological conditions. Highest concentrations occur along major traffi arteries during morning and evening rush hours and decrease relatively rapidly with distance from roadways. Peak levels averaged over 1 h vary from about 5 to more than 10 ppmv in some cities. Elevated CO levels tend to plague cities that have high traffi densities and are located at relatively high altitudes.

CO concentrations are significantly and more important than regulatory concerns would indicate. Carbon monoxide affects tropospheric concentrations of both hydroxyl radical (OH-) and O3 and, as a consequence, the oxidizing potential of the atmosphere. Because CO and CH_4 (a major greenhouse gas) compete for OH-, CO indirectly affects tropospheric concentrations of CH_4 and stratospheric H_2O vapor (derived from CH_4 oxidation).

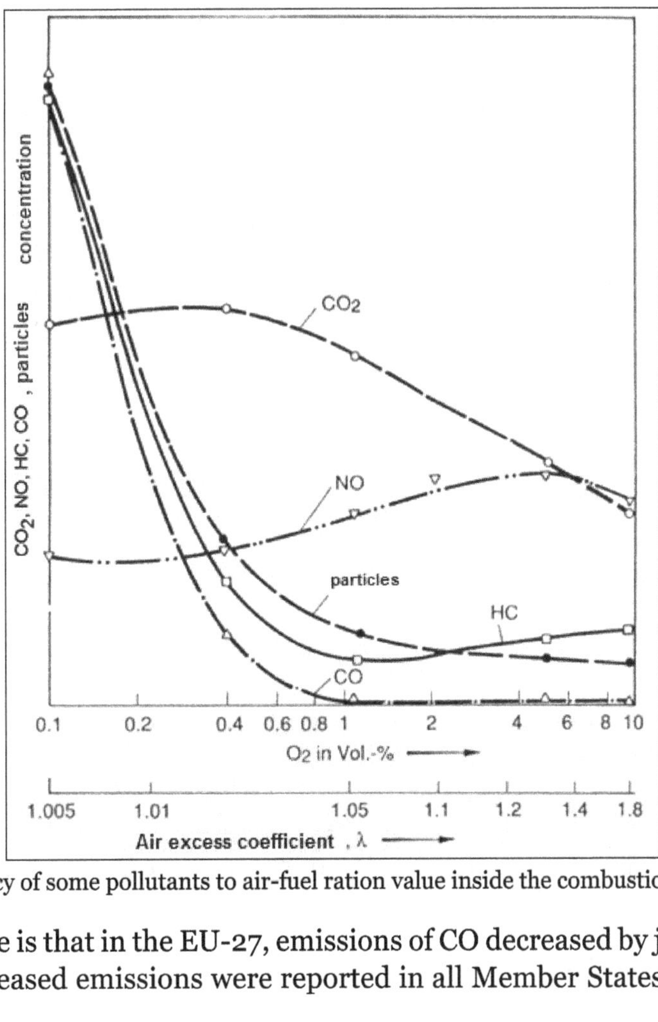

Dependency of some pollutants to air-fuel ration value inside the combustion chamber.

A very important issue is that in the EU-27, emissions of CO decreased by just over 53 % between 1990 and 2006. Decreased emissions were reported in all Member States except Romania. The largest absolute decreases were reported by France, Germany, Italy, Poland, Spain and the United Kingdom; however, these countries remained the largest emitters of CO in absolute terms in 2006.

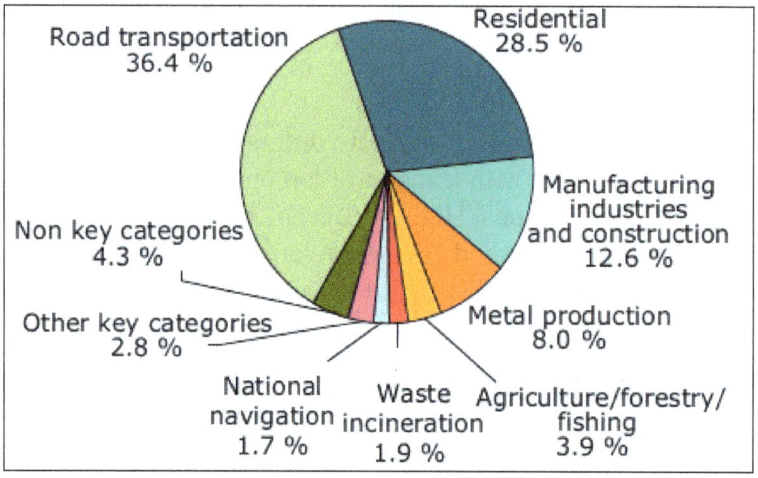

EU-27 emission sources of carbon monoxide.

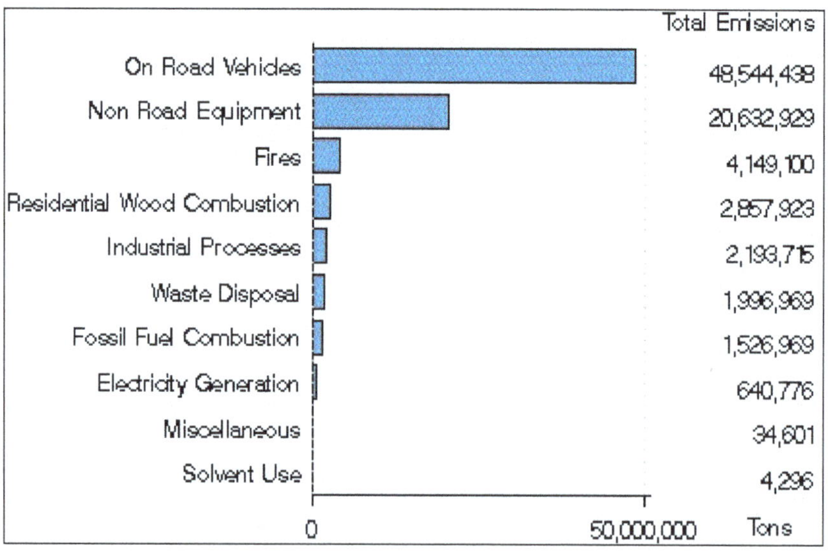

USA emission sources of carbon monoxide.

Carbon monoxide is oxidized in the atmosphere to CO_2 on reaction with OH- radicals:

$$CO + OH^- \rightarrow CO_2 + H^-$$

Subsequent reactions of H- and oxygen O_2 produce the highly reactive hydroperoxyl radical, HO_2-:

$$H^- + O_2 + M \rightarrow HO_2^- + M$$

Where, M is an energy-absorbing molecule, e.g., nitrogen N_2 or O_2. From those two reactions we can conclude:

$$CO + H^- + O_2 \rightarrow CO_2 + HO_2^-$$

In the presence of photons of light energy another reaction will occur:

$$CO + O_2 + hv \rightarrow CO_2 + O_3$$

Where, O_2 is ground-state atomic oxygen and *hv* is a photon of light energy.

As a result of these reactions, CO is converted to CO_2, with one molecule of O_3 produced for each CO molecule oxidized. As such, CO oxidation in the troposphere is an important (but low-level) source of O_3.

The effects of carbon monoxide on human health are caused due to the fact that CO enters the bloodstream through the lungs and attaches to hemoglobin (Hb), the body's oxygen carrier, forming carboxy-hemoglobin (COHb) and thereby reducing oxygen (O_2) delivery to the body's organs and tissues. High COHb concentrations are poisonous. Central nervous system (CNS) effects in individuals suffering acute CO poisoning cover a wide range, depending on severity of exposure: headache, dizziness, weakness, nausea, vomiting, disorientation, confusion, collapse, and coma. At lower concentrations, CNS effects include reduction in visual perception, manual dexterity, learning, driving performance, and attention level. Earlier work is frequently cited to justify the statement that CO exposure sufficien to produce COHb levels of ca. 5% would be sufficien to produce visual sensitivity reduction and various neurobehavioral performance deficits. However, the best estimate was that (COHb) would have to rise to 15-20% before a 10% reduction in any behavioral or visual measurement could be observed.

Carbon monoxide (CO) may be the cause of more than one-half of the fatal poisonings reported in many countries: fatal cases also are grossly under-reported or misdiagnosed by medical professionals. Therefore, the precise number of individuals who have suffered from CO intoxication is not known.

CO poisoning occurs frequently; has severe consequences, including immediate death involves complications and often is overlooked. Efforts in prevention and in public and medical education should be encouraged.

Nitrogen Oxides (NO_x)

Nitrous oxide concentrations in the atmosphere have been increasing steadily since preindustrial times, from ~280 to ~320 ppbv today. Estimated annual emissions to the atmosphere are 13.8 Tg (tetragrams) N/year, with ~70% being produced by nitrification and denitrification processes in undisturbed terrestrial environments and the world's oceans. About 3 Tg N/year, or ~8%, is associated with agricultural tillage, fertilizer use, and animal wastes.

In case of combustion equipment the NO_x are formed during the combustion process at high temperatures by the oxidation of nitrogen content of fuel and combustion air. In a first phase only NO is formed while the NO_2 is mainly formed after the combustion in exhaust process when more O_2 is present and into the atmosphere.

The main anthropogenic sources of nitrogen oxides are road transport and public electricity and heat sector.

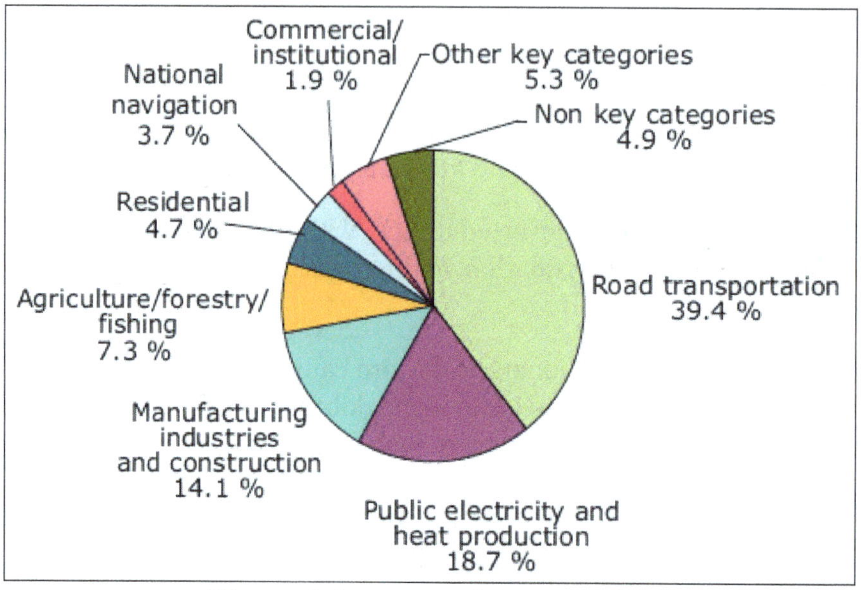

EU-27 emission sources of nitrogen oxides.

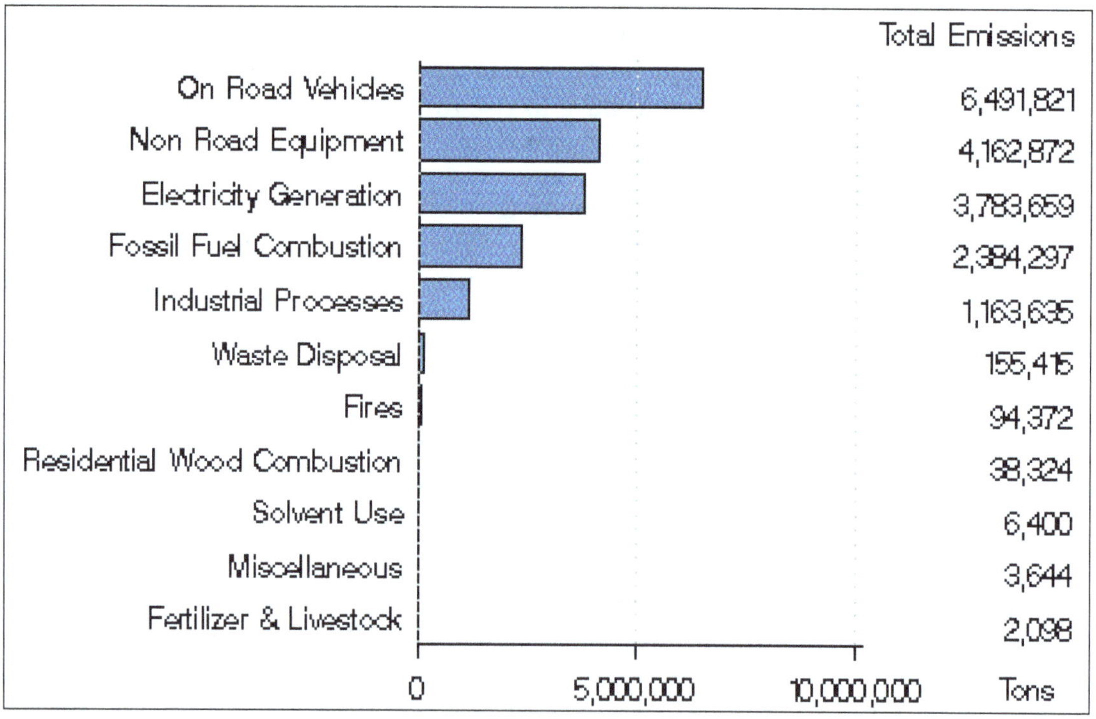

USA emission sources of nitrogen oxides.

Between 1990 and 2006, NO_x emissions decreased in the EU-27 by 35 %. The change of total NO_x emissions between 2005 and 2006 was rather small, a decrease of 1.8 %, mainly caused by reductions achieved in Germany, Italy and the United Kingdom.

Increased atmospheric N_2O levels pose two major environmental concerns: stratospheric O_3 depletion and, because of its thermal absorptivity, global warming. In case of combustions formed nitric oxides three mechanism are present: Thermal mechanism, the prompt mechanism and NO_x formation from fuel nitrogen.

Thermal Mechanism of Nitrogen Oxide Formation

The thermal mechanism of nitric oxide formation during combustion was first studied by Zeldovici. The thermal mechanism occurs in combustion chamber in areas with higher oxygen concentration or in exhaust gases when nitrogen reacts with free oxygen atoms.

In areas when the combustion flame has high oxygen content the following reactions take place:

$$O + N_2 \rightarrow NO + N$$

$$N + O_2 \rightarrow NO + O$$

In areas with rich fuel content and at temperatures over 1300 ◦C another reaction is more probable:

$$N + OH \rightarrow NO + H$$

The quantities of nitric oxides produced through this mechanism are influenced by numerous factor but most relevant are:

- Air-fuel excess ration factor (λ).
- Combustion temperature.
- Air-fuel mixture residence time in combustion chamber areas with high temperatures.

Figure gives an image on how temperature and residence time influences the NO formation during combustion process.

In the case of combustion in steam generators from large power plants, in installation where the ash is evacuated in dry state (particles) the contribution of thermal NO_x mechanism to total NO_x formation is up to 20%. For installations where ash is evacuated in liquid state the thermal NO_x contribution is up to 50%.

In case of piston engines, gas turbines or in any other combustion process at high temperature rate the thermal NO_x mechanism is responsible for ~ 50% of total concentrations of NO_x exhausted. In these cases a very important role plays the air-fuel ratio (λ). If air-fuel ration is lower than 1 the formation of NO decreases.

Prompt NO Mechanism

In areas where the combustion take place with low oxygen content nitric oxide can be formed by reactions between molecular nitrogen N_2 and radicals like CH and CN. This phenomenon was first discovered by Fenimore. It is widely accepted that reactions described below are responsible for NO formation by prompt NO_x mechanism. The majority of scientists agree that the prompt mechanism of NO formation can be neglected because the atomic nitrogen and CN radicals can only be formed in areas with very low air-fuel ratios.

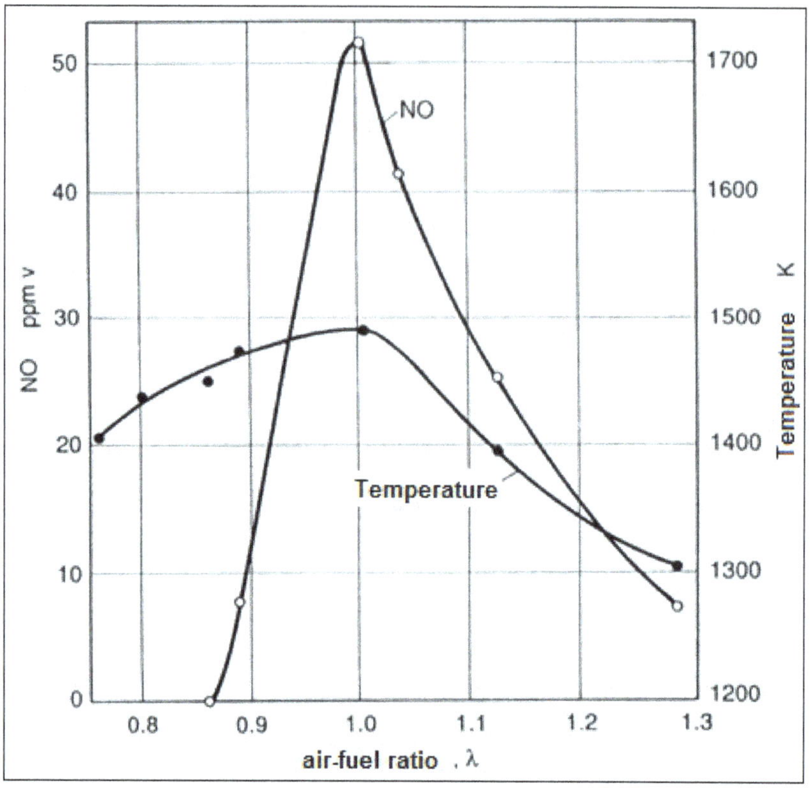

Dependency between formed NO and combustion temperature and air-fuel ration (λ).

$$CH + N_2 \rightarrow HCN + N$$

$$C + N \rightarrow CN + N$$

$$CN + H_2 \rightarrow HCN + H$$

$$CN + H_2O \rightarrow HCN + OH$$

$$HCN(CN) + O \rightarrow NO + R$$

Where, R is an organic residue.

Fuel NO Mechanism

The fuel NO_x mechanism is also a flame based one but is more complex then prompt and thermal NO_x mechanisms and partially unknown. However, it is known that first CN radicals are formed and their evolution in the presence of oxygen leads to NO formation. The decisive factors in fuel NO_x mechanism are nitrogen content of fuel, oxygen concentration in combustion flame, reaction time and flame temperature. The thermal NO_x mechanism is much slower than prompt mechanism and the fuel NO_x mechanism is faster than thermal NO_x.

The sources of nitrogen fuel content are mainly the decay of proteins and nucleic acid from fossil material, mainly the aliphatic compounds as first, secondary and tertiary amines (RNH_2, R_1R_2NH, $R_1R_2R_3N$) or aromatic compound like pyridine (C_5H_5N).

The complicated mechanism of fuel NO_x formation can be synthesized as follows:

$$fuelN \rightarrow HCN \rightarrow NCO \rightarrow \begin{matrix} NH_i \\ N \end{matrix} \begin{matrix} \rightarrow NO \\ \rightarrow N_2 \\ \rightarrow NO \\ \rightarrow N_2 \end{matrix}$$

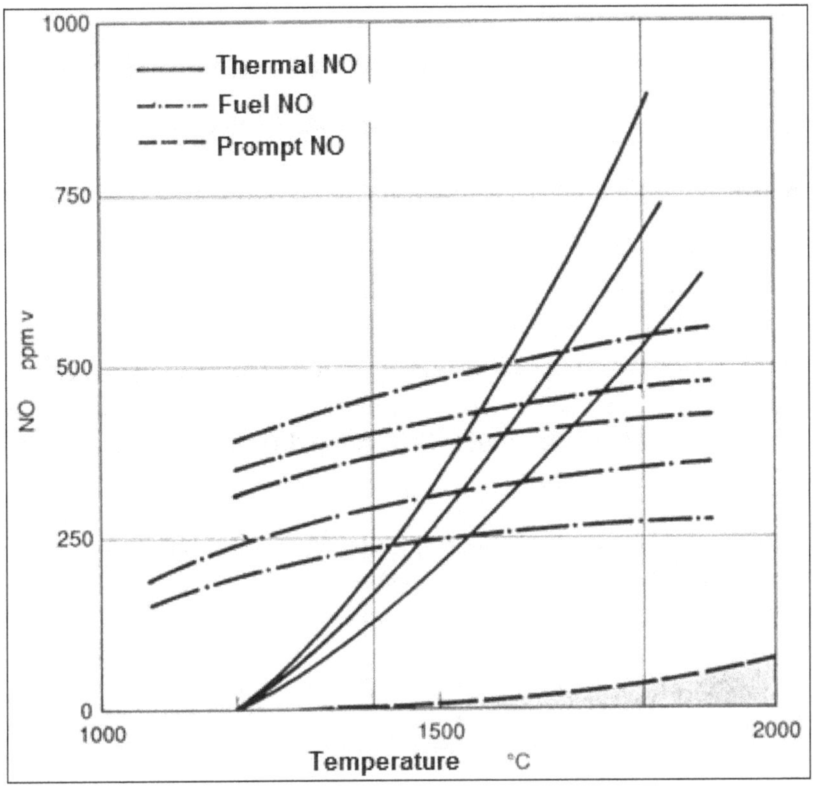

Dependency between NO formation mechanisms and combustion temperature.

Into exhaust gases the nitric oxides consists only in nitrogen oxide (NO). The oxidation of NO to NO_2 occurs partially in flue gases exhaust channels and in atmosphere, most important reactions are:

$$2NO + O_2 \rightarrow 2NO_2$$

$$NO + O_3 \rightarrow NO_2 + O_2$$

Another important reaction is the one that oxidize NO_2 into nitric acid HNO_3 in gaseous phase, because is a major contributor to acid rain by its diffusion in water droplets

$$NO_2 + OH + M \rightarrow HNO_3 + M$$

Where, M is a passive reaction partner.

Sulphur Oxides (SO_x)

All fossil fuels contain sulphur, most of which is released as sulphur dioxide during combustion.

Almost all the anthropogenic sulphur contribution is due to fossil fuel combustion. Different fuels offer a wide range of sulphur contents

- Oil and its by-products contain between 0.1% sulphur (paraffin and 3% (heavy fuel oil) in the form of sulphides and thiols. Petrol contains negligible sulphur in the context of overall mass emissions, although there can be an odor problem from conversion to hydrogen sulphide (H_2S) on catalytic converters.
- Coal contains 0.1–4% sulphur, mainly as flakes of iron pyrites (FeS_2). The average sulphur content of European coal reservoirs is 1.7%.
- Natural gas (mainly methane, CH_4) can be up to 40% H_2S when it is extracted from the well. The sulphur is taken out very efficientl at a chemical processing plant before distribution, so natural gas is effectively sulphur free

Global sulphur dioxide emissions are estimated to have increased from 4 Mt (containing 2 Mt of sulphur) in 1860 to 150 Mt in 1990. The emissions from the US and Europe increased steadily until the 1970s before coming under control. Sulphur emissions from the faster-growing Asian region have continued to increase, due largely to coal combustion. Emissions from China are now comparable to those from the US, and in 1990 emissions from China, US and Russia accounted for over half the global total.

The EU-27 total SO_X emissions in 2006 were 7 946 Gg. This is almost a 70 % reduction compared to the level of emissions reported in 1990. Since 1990, SO_X emissions have increased in only two Member States: Romania (21.9 %) and Greece (11.9 %). Inspection of the time-series trends for some Member States shows some step changes in emission reductions have occurred since 1990. For example, emissions of SO_2 in Slovenia fell considerably in 2001 and again in 2005 due to the introduction of flue gas desulphurization abatement equipment in thermal power plants.

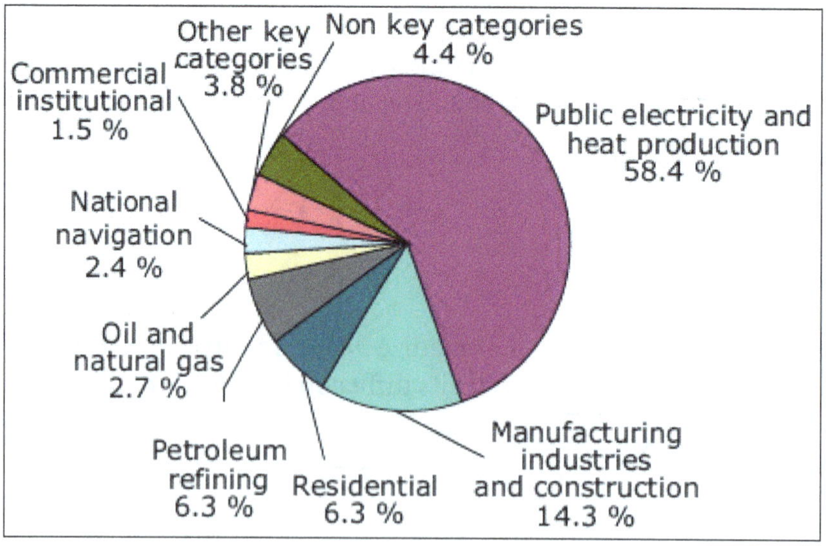

EU-27 emission sources of sulphur oxides.

The natural sources are now heavily outweighed by human ones, principally fossil fuel combustion. Since 90% of the biogenic emission is as dimethyl sulphide, and an even higher proportion of the human emission is as SO_2, we have a clear demarcation between the source types. Since most of the dimethyl sulphide comes from oceans in the southern hemisphere, and most of the human SO_2 from fossil fuel emissions in the northern hemisphere, we also have a geographical split.

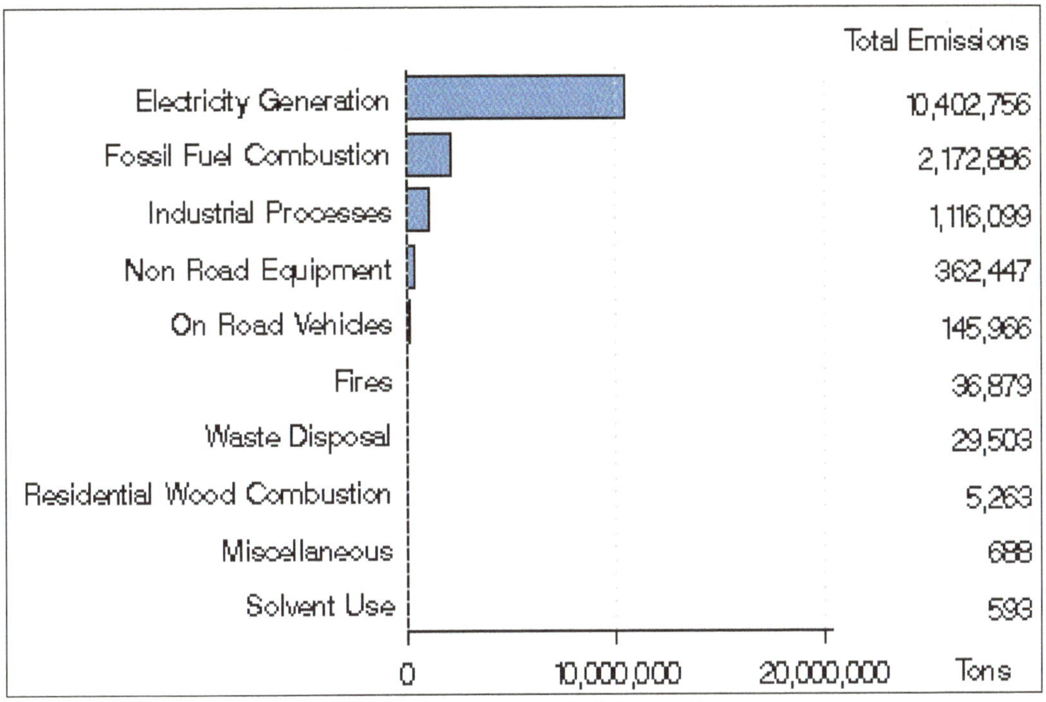

USA emission sources of sulphur oxides.

Anthropogenic SO_2 is almost exclusively formed during the combustion of fossil fuels with relevant sulphur content and most important is coal because the sulfur content of liquid and gaseous fossil fuels can be easily removed before combustion.

If the sulphur content of the fuel is totally combusted SO_2 is formed through the reaction:

$$CH_3 - SH + 3O_2 \rightarrow SO_2 + CO_2 + 2H_2O$$

During the incomplete combustion with lack of oxygen elementary sulphur and hydrogen sulphide (H_2S) can form, at high temperatures:

$$CH_3 - SH + 0.5O_2 \rightarrow H_2S + HCHO$$

$$2H_2S + O_2 \rightarrow 2H_2O + 2S$$

Hydrogen sulphide (H_2S) can occur during the combustion process of low quality coals like lignite or into the piston engines exhaust gases after catalytic reduction if the engine works at high load and low air-fuel ratio.

During the combustion and in exhaust channels the SO_2 can be oxidized to SO_3 who will form with water content of flue gases sulphuric acid H_2SO_4. The SO_2 conversion to SO_3 is increased in the presence of vanadium of iron oxides at temperatures over 800 °C.

Evacuated into the atmosphere SO_2 reacts in proportion of (1 – 2) ‰/hour with oxygen, under the presence of ultraviolet radiation (uvr) and will form SO_3:

$$2SO_2 + O_2 + uvr \rightarrow 2SO_3$$

Afterwards, the SO_3 will react with water vapors from the atmosphere to form sulphuric acid H_2SO_4. In periods of fog or days with high humidity the transformation rate of SO_3 to H_2SO_4 can be up to 15 %.

$$SO_3 + H_2O \rightarrow H_2SO_4$$

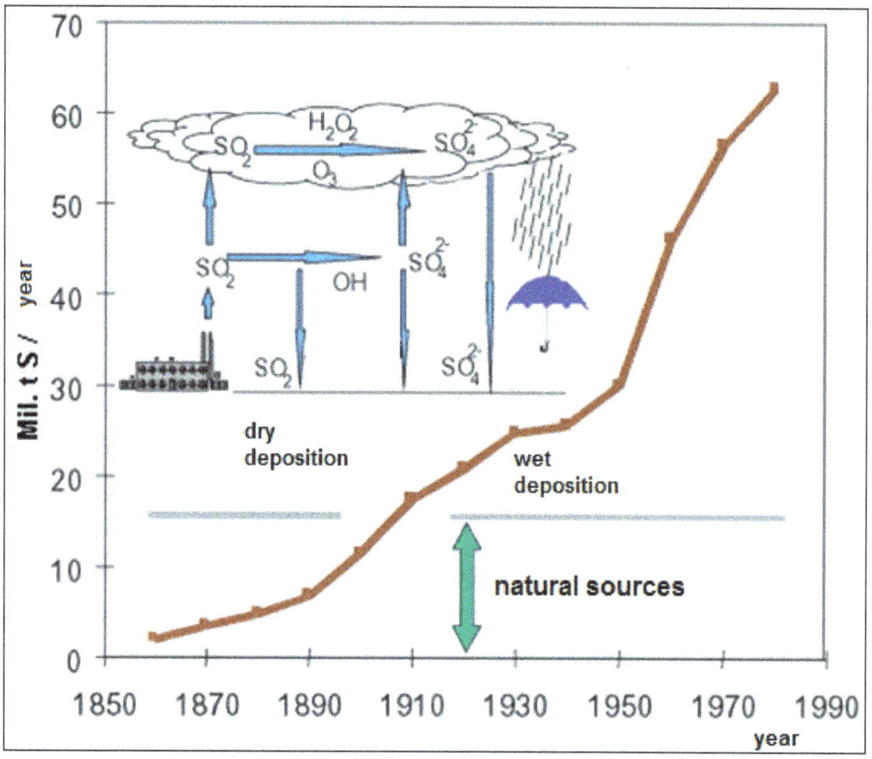

SO_2 atmospheric reaction and historical emissions.

Particulate Matter

Particulate matter is a collective term used to describe small solid and liquid particles that are present in the atmosphere over relatively brief (minutes) to extended periods of time (days to weeks). Individual particles vary in size, geometry, mass, concentration, chemical composition, and physical properties.

They may be produced naturally or as a direct or indirect result of human activities. Of major concern are particles < 10 µm because they can remain suspended in the atmosphere where (depending on actual particle size) they can settle out relatively slowly. Atmospheric aerosol is characterized by these relatively small particles. The formation and increase of atmospheric aerosol by both natural and anthropogenic sources is a major air quality concern because aerosol particles may scatter light, reducing visibility, pose an inhalation hazard to humans and animals, affect climate on a regional and global scale.

Particulate matter can be described as being primary or secondary, based on its origin and processes of formation. Primary particles are emitted directly into the atmosphere from a variety of natural and anthropogenic sources. The natural particle sources include volcanoes, forest fires, ocean spray, biologic sources and the anthropogenic sources of particles are transportation, fuel combustion in stationary sources, a variety of industrial processes, solid waste disposal and miscellaneous sources such as agricultural activities and fugitive emissions from roadways. Secondary particles are formed in the atmosphere as a result of chemical processes involving gases, aerosol particles and moisture.

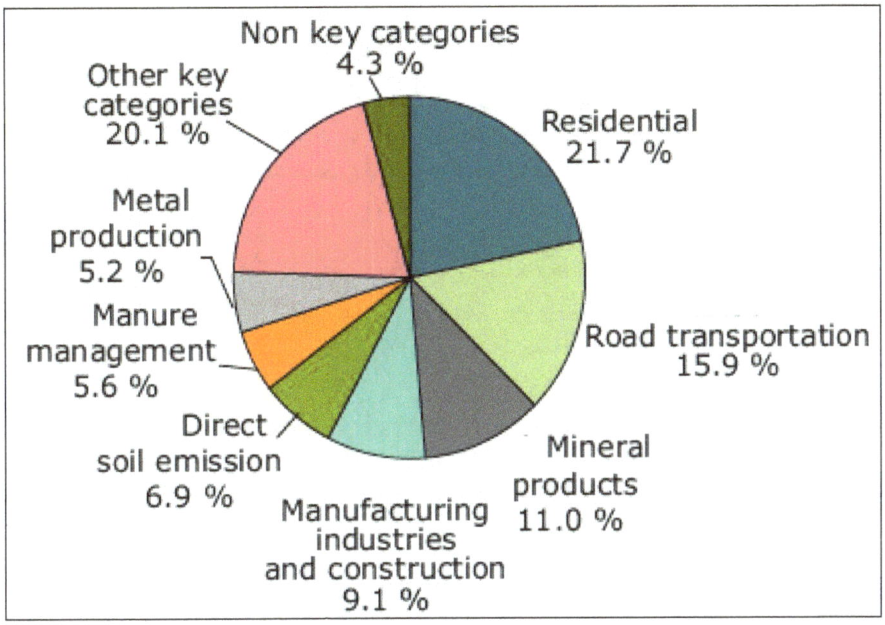

EU-27 emission sources of PM10.

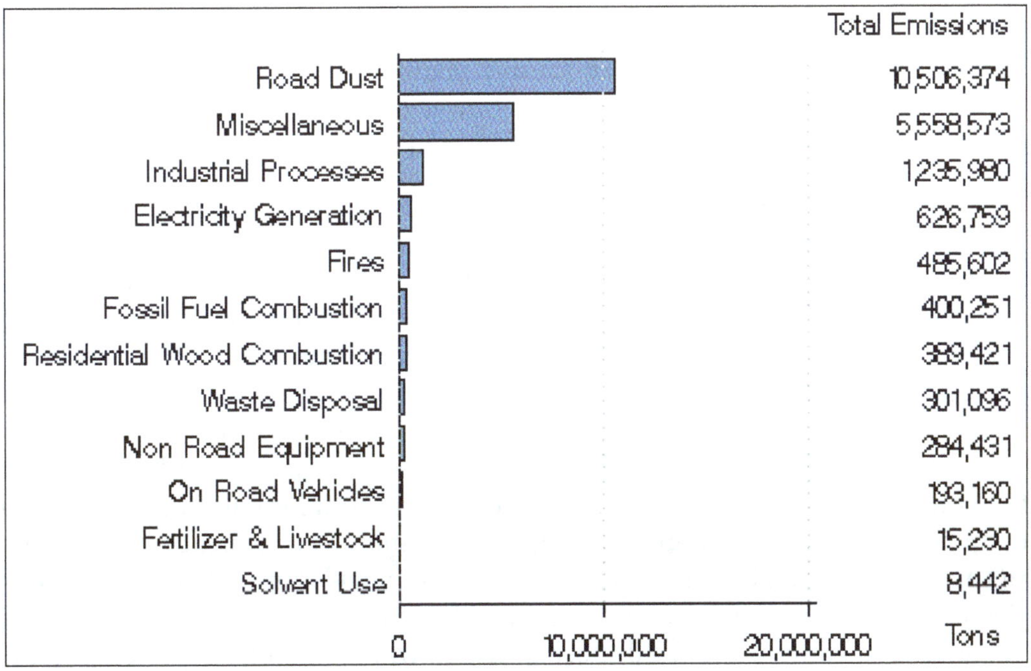

USA emission sources of PM10.

In Figure below the sources and diameter of representative particle matter fractions, thoracic (PM10), Respirable (PM4.0) and total suspended particles (TSP) are synthesized:

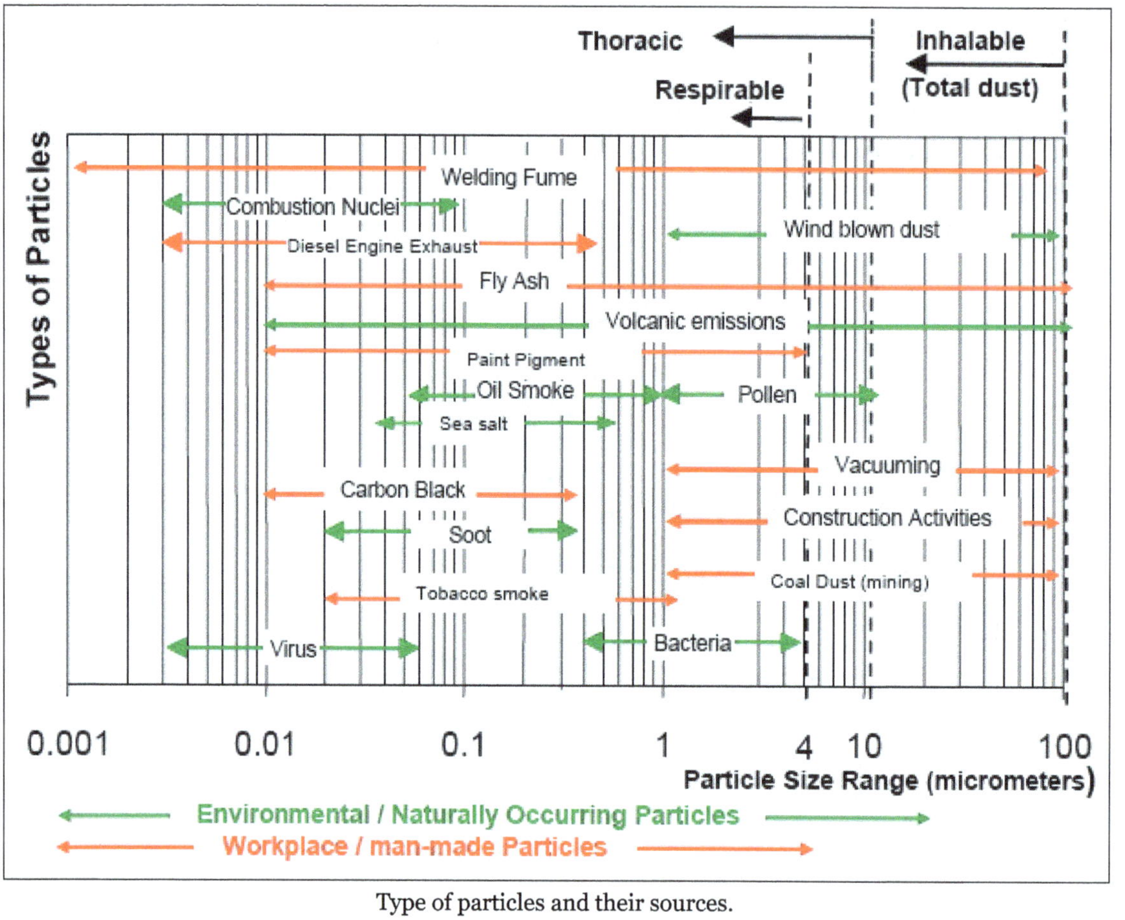

Type of particles and their sources.

Into the combustion process particles are formed mainly from agglomerations of carbon and partially from hydrocarbons. The particle formation is encouraged by lack of oxygen at flam base and by high temperatures in this phase. The type of fuel is also relevant for particle formation.

During the fuel combustion the particles can form as soot, condensed hydrocarbons, ash or unburned carbon. For example, in the case of coal combustion, when the process starts firs the volatile compounds is released while the mineral components are almost completely unburned. Large size particles (from 1 to 20 μm) are formed by mineral inclusions agglomeration on burned soot. During the soot fragmentation up to 5 large ash particles are formed, with a diameter from 10 to 20 μm and a large number of particles with a diameter of 1 to 3 μm. Particles with a diameter lower then 0.1 μm will form mainly due to vaporization and re-condensation of a part of the mineral components that are surrounding the ignited coal. In the combustion conditions of up to 1400 to 1600 °C appreciatively 1 % of mineral components will vaporize in Na, As, Fe, SiO and Mg vapors. This vapors will then diffuse (nucleation) into combustion chamber extremities where oxygen content is higher and will form a large number of very fine particle with a diameter of maximum 1 μm. All those particles will be then transported in atmosphere by exhaust gases.

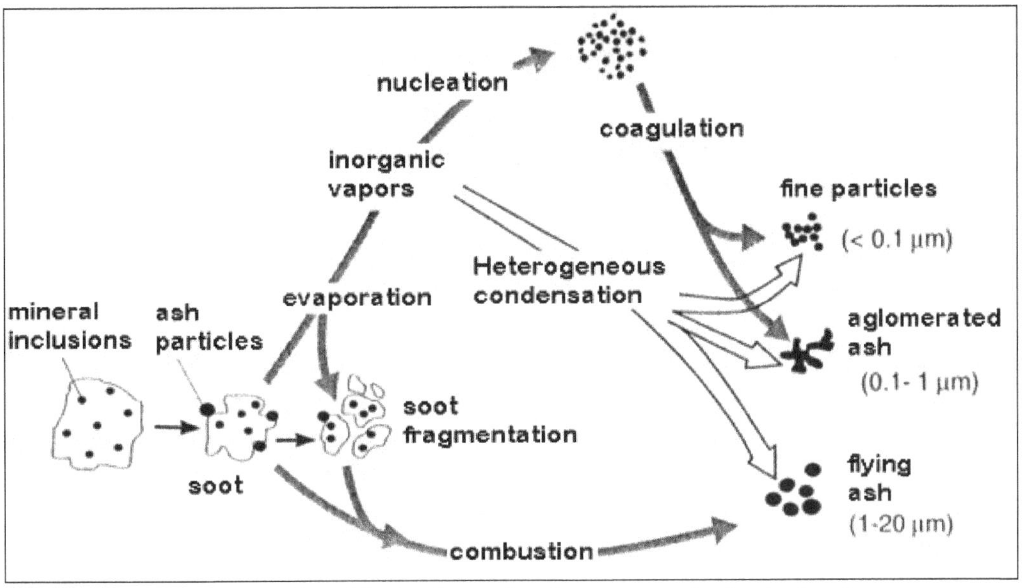

Formation of particles in the combustion process of pulverized coal.

Before atmospheric particles can cause adverse health effects, they must enter and be deposited in the human respiratory system. Respiratory penetration and deposition depend on the aerodynamic size of particles, respiratory defense mechanisms, and breathing patterns. Particles can be described as inhalable, thoracic or respirable, based on their penetration and potential for deposition. Particles with a diameter higher than 10 μm generally do not pass through the nasal hairs and defense mechanisms of the upper respiratory system and are not of a public health concern.

Non-Methane Volatile Organic Compounds

The anthropogenic fraction of atmospheric VOCs is related to the unprecedented usage of fossil fuels for transport, the production of consumer goods and various industrial processes in the past centuries. The distinction between biogenic and anthropogenic VOCs in the atmosphere is far from straightforward because many VOC species are produced by both sources. Emissions of alkanes and alkenes, for example, are dominated by anthropogenic sources, but are also produced by soils, wetlands and oceans.

The large sources of NMVOC emissions are use of fossil fuel in transportation and chemistry industry. Mobile sources can be divided into emissions from the exhaust and fugitive emissions by evaporation. Stationary emissions from the use of fossil fuel are due to industrial applications (e.g. refineries and chemical sector). Emissions related to production, storage and delivery of fossil fuels predominately occur in those regions where extensive fossil fuel drilling activities exist. However, fugitive emissions can also occur from the transport and distribution of the fuel, such as ships, road tankers and fuel stations.

After their release into the atmosphere, VOCs are oxygenated by photochemical processes, which finally lead to their removal from the atmosphere. For most VOCs the process is initiated by atmospheric radicals like OH, O_3, NO_3 and Cl, with the OH radical being by far the most important reactant. The atmospheric lifetime of an individual VOC species is dependent on its chemical structure, the radical concentration and the intensity of solar radiation. When VOCs are degraded in polluted air masses, NO is oxygenated to NO_2, which then gets photolysed and contributes to the formation of tropospheric ozone, a key issue in air pollution control.

In the EU-27, NMVOC emissions declined by just under 45 % between 1990 and 2006. Twenty-three countries reported reductions (Belgium, Germany, Luxembourg the Netherlands and the United Kingdom have reduced emissions by more than 60 % during this period). The four countries that reported increased NMVOC emissions are Bulgaria, Greece, Poland and Romania.

The ability of NMVOCs to cause health effects varies greatly from those that are highly toxic, to those with no known health effect. As with other pollutants, the extent and nature of the health effect will depend on many factors including level of exposure and length of time exposed. Eye and respiratory tract irritation, headaches, dizziness, visual disorders, and memory impairment are among the immediate symptoms that some people have experienced soon after exposure to some organics. At present, not much is known about what health effects occur from the levels of organics usually found in homes. Many organic compounds are known to cause cancer in animals; some are suspected of causing, or are known to cause, cancer in humans.

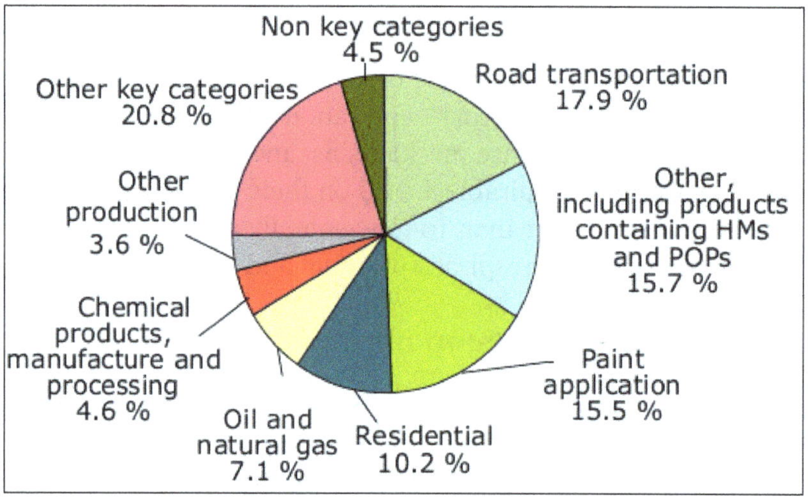

EU-27 emission sources of NMVOC.

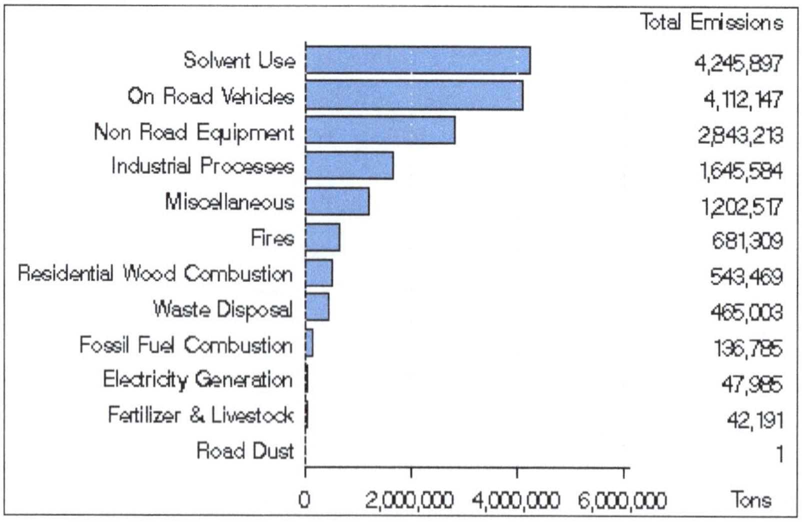

USA emission sources of NMVOC.

Relevant Methods to Control NO_x and SO_x Emissions from Fossil Fuel Combustion

NO_x control

Two primary categories of control techniques for NO_x emissions are combustion control and flue gas treatment. Very often more than one control technique is used in combination to achieve desired NO_x emission levels. A variety of combustion control techniques are used to reduce NO_x emissions by taking advantage of the thermodynamic and kinetic processes. Some reduce the peak flame temperature; other reduces the oxygen concentration in the primary flame zone while other methods use the thermodynamic balance to reconvert NO_x back to nitrogen and oxygen.

In the low air-fuel excess ration firing techniques the principle is based on cutting back the amount of excess air, the lower oxygen concentration in the flame zone reduces NO_x production. In some cases where too much excess air has become normal practice, thermal efficienc is improved. However, low excess air in the resulting flame may be longer and less stable, and carbon monoxide emissions may increase. Applying advanced optimization systems at four coal-fired power plants resulted in NO_x emission reductions of 15 to 55%.

Another widely used method to control NO_x emissions is the flue gas recirculation technology, when some of the flue gas, which is depleted in oxygen, is re-circulated to the combustion air. This has two effec s: the oxygen concentration in the primary flame zone is decreased, and additional nitrogen absorbs heat, and reduces the peak flame temperature. Injecting water or steam into the combustion chamber provides a heat sink that reduces peak flame temperature.

Low-NO_x burners are designed to stage either the air or the fuel within the burner tip. The principle is similar to overfire air (staged air) or reburn (staged fuel) in a furnace. With staged-air burners, the primary flame is burned fuel rich and the low oxygen concentration minimizes NO_x formation. Additional air is introduced outside of the primary flame where the temperature is lower, thereby keeping the thermodynamic equilibrium NO_x concentration low, but hot enough to complete combustion. Staged-fuel burners introduce fuel in two locations. A portion of the fuel is mixed with all of the combustion air in the first zone, forming a hot primary flame with abundant excess air. NO_x formation is high in this zone. Then additional fuel is introduced outside of the primary flame zone, forming a low-oxygen zone that is still hot enough for kinetics to bring the NO_x concentration to equilibrium in a short period of time. In this zone, NO_x formed in the primary flame zone reverts back to nitrogen and oxygen.

All those methods are primary methods to reduce NO_x formation at the combustion chamber level. Also secondary methods for NO_x reduction have been developed, like selective non-catalytic reduction (SNRC) and selective catalytic reduction (SCR). Selective non-catalytic reduction uses ammonia (NH_3) or urea (H_2NCONH_2) to reduce NO_x to nitrogen and water. The overall reactions using ammonia as the reagent are:

$$2H_2S + O_2 \rightarrow 2H_2O + 2S$$

$$8NH_3 + 6NO_2 \rightarrow 7N_2 + 12H_2O$$

The intermediate steps involve amine (NHi) and cyanuric nitrogen (HNCO) radicals.

The critical dependence of temperature requires excellent knowledge of the temperature profile within the furnace for placement of reagent injection nozzles. In the case that the SNCR process is not controlled efficiently supplementary emissions will occur in exhaust gases, like CO, NH_3 or N_2O, called secondary emissions. In a typical application, SNCR produces about 30 to 50% NO_x reduction. Some facilities that require higher levels of NO_x reduction take advantage of the low capital cost of the SNCR system, then follow the SNCR section with an SCR system.

In the SCR technology a catalyst bed can be used with ammonia as a reducing agent to promote the reduction reaction and to lower the effective temperature. An SCR system consists primarily of an ammonia injection grid and a reactor that contains the catalyst bed. A variety of catalyst types are used for SCR. Precious metals are used in the low temperature ranges. Vanadium pentoxide on titanium dioxide is a common catalyst for the medium temperature ranges and various aluminum silicates are used as high temperature catalysts. While the SNCR technology can provide NO_x reduction ratios of over 90% has a major disadvantage in economical cost and the necessity to retrofit the combustion facilities

SO_2 Control

SO_2 control processes are used for coal-fired industrial boilers. SO_2 and HCl controls are required for hazardous and municipal solid waste combustors. Many coal fired power plants use wet limestone scrubbers that have a relatively high capital cost in order to utilize inexpensive limestone reagent. Smaller, industrial-scale facilities typically use more expensive reagents in systems with lower equipment costs. Another solution to control SO_2 emissions can be found in combustion of coals (anthracite) with high calorific value and low sulphur content.

The most relevant and used technology to reduce SO_2 emissions from coal fired power plants is the combustion of coal with calcium based absorbers. Limestone is an inexpensive rock that is quarried and crushed. It can be used directly as a reagent either in aqueous slurry or by injection into a furnace where the heat de-carbonates the limestone. This is a primary reduction process. The parameters that will influence the efficienc of SO_2 removal are the type of combustor, the type of coal, the absorber quality and its time of residence into the facility.

The main reactions involved are:

$$CaCO_3 \xrightarrow{\sim 800 grd.C} CaO + CO_2$$

$$CaO + SO_2 \rightarrow CaSO_3$$

$$CaSo_3 + \frac{1}{2}O_2 \rightarrow CaSO_4$$

Other absorbers can also be used, like $CaOH_2$, Na_2CO_3 or NaOH but they are more expensive than limestone. The secondary methods for SO_2 reduction can be classified as dry procedures, semidry and wet procedures. A simplified process flow diagram for a coal fired power plant with wet limestone SO_2 emission control system is presented in Figure.

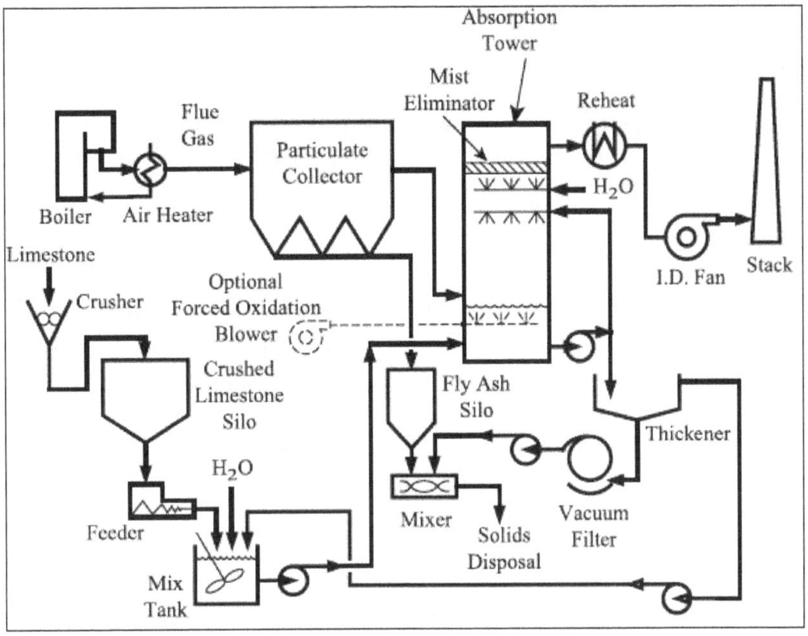

Simplified wet limestone process flow diagra

Indoor Air Pollution

When we think of air pollution, we probably think of things like smog, power plant emission and emissions from vehicles. These are all examples of outdoor air pollution, but indoor air can be polluted too. Pollutants are any harmful contaminants in the air; therefore, indoor air, which is closely connected with outdoor air cannot escape from the effects of outdoor contamination. Even things or activities present indoor are also contribute largely in indoor air pollution. Indoor air pollution is a very real and dangerous thing because indoor air is far more concentrated with pollutants than outdoor air. We also spend most of the time inside the building in comparison to outdoor activities. There are many sources of indoor air pollution, but they are different for city and villages, commercial region to industrial region, and developed & developing nations.

Indoor air pollution refers to chemical, biological and physical contamination of indoor air. It may result in adverse health effects. The Indoor air quality is determined by the concentrations of pollutants and thermal (Temperature and Relative Humidity) conditions that has impact on the comfort and performance of occupants. The air indoors presents a more complex mixture of these contaminants in a localized area than outdoor environment. The four elements: source, HVAC system, pollutant pathways and occupants are involved in development of IAQ(Indoor Air Quality) problems. The air quality status of a building/ enclosed space is therefore directly associated with the health of its occupants. A number of factors associated with indoor air have its impact on quality. These may include:

- Inadequate ventilation and poorly designed ventilation systems.
- Flow of outdoor air contaminants into the internal spaces.

- Extreme of temperature and humidity levels.
- Combustion activities.
- Making use of cleaning products, paints, printers, pesticides and other VOCs generating products etc.

Indoor Air Pollutants

Many indoor pollutants are either known, or suspected to be, allergens, carcinogens, neurotoxins, immune-toxins or irritants, while all may contribute to sick building syndrome (SBS). The IAQ and the presence of air pollutants in indoor environment is a worldwide issue, since many governments and environmental institutes have faced this serious phenomenon. There are many indoor air contaminants, which can be separated based on their effects on human health, the frequency of their appearance, their usual concentration levels, their sources etc.

Carbon dioxide from people, pets, and plants:

- Living things breathe out carbon dioxide.
- A high level of carbon dioxide in the indoor means there is an inadequate supply of fresh air. This is usually caused by:
    - Improper partitioning and overcrowding.
    - Infrequent window opening.
    - Improper maintenance or operation of the ventilation system.
- It will make you feel sleepy, and provide a warning for possible build-up of other indoor air pollutants.

Biological contaminants include bacteria, fungi (commonly known as mould), viruses and dust mites:

- Their growth may be accelerated by:
    - Inadequate ventilation.
    - Damp and wet condition.
    - Dirty environment.
    - Dirty air filter and ducting of air conditioners or ventilation systems
- They may cause sneezing, watery eyes, coughing, shortness of breath, dizziness, lethargy, fever, digestive problems.
- Some of them may trigger allergic or asthmatic reactions.

Radon from building materials:

- Radon is a colourless and odourless radioactive gas that comes out from concrete building materials containing granite.

- It can accumulate to a high concentration if the premises are poorly ventilated.
- Exposure to elevated radon and its decay products may increase the incidence of lung cancer.

Volatile Organic Compounds (VOCs) from many sources such as building materials, furniture, cosmetics, fragrance products, cleaning agents, pesticides, adhesives, paints, dry-cleaned clothing and tobacco smoke:

- There are over a hundred compounds, including formaldehyde, that can be classified as VOCs in a typical indoor environment.
- They may cause discomfort, and even worse, may cause cancer.

Formaldehyde emissions indoor mainly come from furniture made of wood products:

- Formaldehyde is mainly emitted from the urea-formaldehyde resins used in the pressed-wood panels (e.g. particle board, medium-density fiberboard and plywood) or the coating materials.
- Other sources include gas cooking, incense burning, carpeting, etc.
- High formaldehyde level will irritate your eyes, nose and throat.

Environmental tobacco smoke (ETS) from smoking (Second hand smoking):

- ETS is a mixture of the smoke given off by the burning end of a cigarette, pipe, or cigar, and the smoke exhaled by the smoker.
- It is a complex mixture of over 4,000 chemical compounds.
- ETS is a strong irritant and a known cancer causing agent. It causes eye, nose, and throat irritation and may significantly increase the risk of lung cancer and other respiratory illnesses.

Ozone is produced by equipment, including photocopiers, laser printers and ionisers, that utilises ultra-violet light or causes ionisation of air:

- Ozone is highly reactive and could cause serious damage to the lung if inhaled in high concentrations.
- It also irritates the eyes and respiratory tract.
- As it is so reactive, it is only found in substantial concentrations near the source, and generally does not tend to accumulate in the indoor environment.

Airborne particles drawn into the building from outside and from indoor activities, like cooking, operating equipment and smoking. These light weight airborne particles may cause eye, nose and respiratory irritation.

Indoor Air Pollutant	Source	Potential Health Effect	Measures to Improve IAQ
Carbon dioxide (CO_2)	• Human breathing. • Tobacco smoking. • Combustion sources.	No significant health impact in general but causes sleepiness, and even headaches, dizziness and nausea at high level.	Provide sufficient fresh ai supply and ventilation.

Carbon monoxide (CO)	• Combustion sources like cooking stove. • Vehicle/engine exhaust. • Tobacco smoking.	Causes headaches, nausea, fatigue, rapid breathing, chest pain, decreased alertness/ confusion/ impaired judgment and is even fatal at high level.	Avoid combustion in indoor environment or provide separate exhaust system, maintain sufficien ventilation, not to idle vehicle inside car park.
Respirable suspended particulates (RSP or PM_{10})	• Combustion sources like cooking stove. • Equipment like photocopiers and printers. • Tobacco smoking. • From outdoors.	Causes irritation to eyes, nose, skin and, respiratory system, and sometimes allergic reactions.	Clean and vacuum frequently, maintain ventilation system properly, use proper air purifier, not to smoke indoor.
Nitrogen dioxide (NO_2)	• Combustion sources like cooking stove. • Vehicle/engine exhaust. • Smoking. • From outdoors.	Causes irritation to eyes and respiratory system.	Avoid combustion in indoor environment or provide separate exhaust system, maintain sufficien ventilation, not to idle vehicle inside car park, review fresh air intake points.
Ozone (O_3)	• Equipment like photocopiers, laser printers and fax machines. • Air purifiers with high voltage discharge components.	Causes irritation to eyes and respiratory system, and chronic respiratory disease.	Isolate ozone emitting equipment and provide exhaust gas system.
Formaldehyde ($HCHO$)	• Pressed-wood furniture and products. • Adhesives. • Paints. • Urea-formaldehyde foam insulation (UFFI). • Tobacco smoke. • Incense burning.	Causes irritation to eyes and respiratory system, and even lung damage at high level; human carcinogen.	Avoid using pressed-wood products, use low emission furniture and renovation materials, maintain sufficient ventilation.
Total volatile organic compounds (TVOC)	• New furniture and furnishings. • Renovation materials like paints and solvents. • Consumer and aerosol products like cleaning agents, disinfectants, pesticides, cosmetics and ragrance products. • Dry-cleaned clothes. • Smoking.	Causes irritation to eyes, nose and throat, and even headaches, dizziness, visual disorder, and many other impairments; some VOCs are human or animal carcinogens.	Use no/low emission renovation materials and furniture, reduce use of VOC-containing consumer products and aerosol products, maintain proper ventilation.

Radon (Rn)	Concrete building materials containing granite.	Causes lung cancer.	Use sealant or wallpaper to cover up concrete materials, maintain sufficien ventilation.
Biological contaminants (bacteria. fungi/mould, viruses, dust mites)	• Dirty air conditioner/ventilation system filter and ducting. • Growth accelerated by inadequate ventilation, and damp and dusty environment.	Causes sneezing, watery eyes, coughing, shortness of breath, dizziness, lethargy, fever, digestive problems; may trigger allergic or asthmatic reactions.	Keep premises clean and dry, maintain good hygiene and good ventilation, clean ventilation system regularly.
Mould	Grows rapidly on organic matter under wet and warm condition.	Causes allergic reactions to mould-sensitised individuals, e.g. headaches, sneezing, runny nose, red eyes, and skin rashes.	Keep premises clean and dry, maintain relative humidity lower than 70%, fix water leakage/seepage problem, clean up mould spots promptly.
Environmental tobacco smoke	Smoking.	Causes respiratory irritation, adverse effects on cardiova - cular system and lung cancer.	Quit smoking.

Various Factors Attributing to Indoor Air Pollution

It can be categorized as follows:

- Location of building: The location of a building can have direct association with the type of indoor pollutants. Thus buildings in close proximity of highways or busy thoroughfares may exhibit high concentration of particulates and other pollutants while those nearby industries may have higher amount of chemical based gaseous pollutants.
- Building Design and maintenance: The improper building design and construction flaw may also contribute to indoor air pollution. The buildings may suffer from poor foundations, roofs, ceilings, windows and door openings promoting the influx of outdoor pollutants. The non-compliance with the updating of HVAC system needed for accommodating any redesigning process may also lead to infiltration of outdoor pollutants such as particulates, vehicle exhaust, parking garage contaminants, etc.
- Renovation Activities: The renovation processes like painting and polishing may cause the distribution of various gaseous pollutants and other by products throughout the building.
- Building Materials: The presence of damp structural surfaces (e.g., walls, ceilings) or non-structural surfaces (e.g., carpets, shades), may be source of different forms of pollutants including dust, dust mites, microbes, particulates etc. contributing to the indoor air pollution.
- Occupant Activities: Building occupants may use a number of cosmetic and daily used products including perfumes or colognes, detergents, soaps etc. which count among the major pollutant sources.

- Local Exhaust Ventilation: These include such regions with insufficien local exhaust ventilation like kitchens, laboratories, maintenance shops, salons, toilet rooms, laundry rooms and other specialized areas.
- Control of odors and contaminants by diluting them with outdoor air if there is appropriate flow of air that mixes them with room air

Thermal Comfort

A number of variables, including personal activity levels, uniformity of temperature, radiant heat gain or loss, and humidity interact to determine whether people are comfortable with the temperature of the indoor air. The activity level, age, and physiology of each person affect the thermal comfort requirements of that individual. Uniformity of temperature is important to comfort. When the heating and cooling needs of rooms within a single zone change at different rates, rooms that are served by a single thermostat may be at different temperatures. Temperature stratification is a common problem caused by convection, the tendency of light, warm air to rise and heavier, cooler air to sink. If air is not properly mixed by the ventilation system, the temperature near the ceiling can be several degrees warmer than at floor level

Some other technical factors like, the kind of activity being performed within its premises, non-compliance with the product (pesticides, disinfectants, products used for cleaning and polishing) usage, gaseous emissions from the combustion sources (from smoking, kitchens, cafeterias and laboratories), and cross contamination coming from other poorly ventilated zones which then diffuses towards neighbouring areas and affects them. The presence of stagnant water relates to the dominance of various types of biological contaminants, with an amount impregnated into materials, exhausts etc. The defective maintenance of humidifiers and refrigeration towers also contribute to the microflora

As regards anthropogenic activity, three main sources of contamination from outdoor environment include: combustion in stationary sources (power stations); Combustion in moving sources (vehicles) and industrial processes. The major contaminants emitted by these sources are carbon monoxide, oxides of sulphur, and oxides of nitrogen, volatile organic compounds (including hydrocarbons), polycyclic aromatic hydrocarbons (PAHs) and particles. The vehicular combustion contributes as a source of carbon monoxide and hydrocarbons along with more harmful oxides of nitrogen. Combustion in stationary sources and industrial processes produces oxides of sulphur. These also generate more than half of the particles emitted into the air by human activity, and industrial processes can be a source of volatile organic compounds. A number of naturally generated contaminants such as particles of volcanic dust, soil and sea salt, and spores and microorganisms are propelled through the air. The contaminants apart of being introduced from outside sometimes follow a path where they are first expelled out of the building and then return inside as intakes from of the air conditioning system among other sources of pollutants indoors include infiltration through the foundations of the building (e.g., radon, fuel vapours, sewer effluvia fertilizers, insecticides and disinfectants).

Indoor Nitrogen Oxides

Nitrogen monoxide (NO) and nitrogen dioxide (NO_2) are gases produced by high-temperature combustion through reactions between nitrogen and oxygen. Both nitrogen in fuel and nitrogen in the air can participate in the reactions. Formation of NO can be followed by further oxidation into NO_2 in the atmosphere. NO_2 is the most toxic of the nitrogen oxides (grouped generically as NO_x) and it is very important from the health point of view. Emissions of nitrogen oxides have generally increased, a fact which has resulted in a lot of studies dealing with nitrogen oxides exposure, particularly over the last 15 years.

Measurements of NO_2 were carried out in residential and industrial premises over 2-day periods both in winter and summer. Personal exposures were found to vary substantially among the 43 respondents, ranging from 10.2 - 87.5 ppb with a mean of 43.7 16 ppb in winter and 23.6 7.8 ppb in summer. The results also showed that the season of the year, house size and outdoor occupation could significantly influence exposure as well as factors such as the type of fuel used and the ambient air quality level. Personal exposure to NO_2 was significantly greater in winter than in summer, and respondents living in smaller houses and/or having outdoor occupations were exposed to higher NO_2 levels.

Formation and Sources of Nitrogen Oxides

Nitrogen compounds form during combustion. In fact, the atmosphere itself is the source of much of the nitrogen leading to the formation of nitrogen oxides (NO_x). Molecular nitrogen (N_2) makes up most of the gases in the earth's atmosphere (79% by volume). Because N_2 is relatively nonreactive under most atmospheric conditions, it seldom enters into chemical reactions, but under pressure and at very high temperatures, it will react with O_2:

$$N_2 + O_2 \rightarrow 2NO$$

Approximately, 90–95% of the nitrogen oxides generated in combustion processes are in the form of nitric oxide (NO), but other nitrogen oxides can form, especially nitrogen dioxide (NO_2), so air pollution experts refer to NO and NO_2 collectively as NO_x. In fact, in the atmosphere the emitted NO is quickly converted photo-chemically into NO_2.

$$2NO + O_2 \rightarrow 2NO_2$$

$$NO_2 + h\upsilon \rightarrow NO + O$$

$$O + O_2 + M \rightarrow O_3 + M$$

$$NO + O_3 \rightarrow NO_2 + O_2$$

Equation $NO_2 + h\upsilon \rightarrow NO + O$ shows the photochemical dissociation of NO_2. Equation $O + O_2 + M \rightarrow O_3 + M$ shows the formation of ozone from the combination of O and molecular O_2 where M is any third-body molecule (principally N_2 and O_2 in the atmosphere). Equation $NO + O_3 \rightarrow NO_2 + O_2$ shows the oxidation of NO by O_3 to form NO_2 and molecular oxygen. These reactions represent a cyclic pathway driven by photons represented by hυ.

Nitric oxide is a colorless, odorless gas and is essentially insoluble in water. Nitrogen dioxide has a pungent acid odor and is somewhat soluble in water. At low temperatures such as those often present in the ambient atmosphere, NO_2 can form the molecule $NO_2–O_2N$ or simply N_2O_4 that consists of two identical simpler NO_2 molecules. This is known as a dimer. The dimer N_2O_4 is distinctly reddish-brown and contributes to the brown haze that is often associated with photochemical smog.

Both NO and NO_2 are harmful and toxic to humans, although atmospheric concentrations of nitrogen oxides are usually well below the concentrations expected to lead to adverse health effects. The low concentrations are due to the moderately rapid reactions that occur when NO and NO_2 are emitted into the atmosphere. Much of the concern about regulating NO_x emissions stems from the wish to suppress the reactions in the atmosphere that generate the highly reactive molecule ozone (O_3). Nitrogen oxides play key roles in important reactants in O_3 formation. Ozone forms photo-chemically (i.e. the reaction is caused or accelerated by light energy) in the lowest level of the atmosphere, known as the troposphere, where people live. Nitrogen dioxide is the principal gas responsible for absorbing the sunlight needed for these photochemical reactions. So, in the presence of sunlight, the NO_2 that forms from NO incrementally stimulates photochemical smog-forming reactions because nitrogen dioxide is very efficien at absorbing sunlight in the ultraviolet portion of its spectrum. This is why ozone episodes are more common during the summer and in areas with ample sunlight.

High temperature/high pressure conditions exist in internal combustion engines, like those in automobiles (known as "mobile sources"), which is why NO_x is one of the major mobile source air pollutants. These conditions of high temperature and pressure can also exist in boilers such as those in power plants, so NO_x is also commonly found in high concentrations leaving fossil fuel power generating stations. In addition to atmospheric nitrogen, other sources exist, particularly the nitrogen in fossil fuels.

The nitrogen oxides generated from atmospheric nitrogen are known as "thermal NO_x" since they form at high temperatures, such as near burner flames in combustion chambers. Nitrogen oxides that form from fuel or feedstock are called "fuel NO_x." A significant fraction of fuel nitrogen remains in the bottom ash or in unburned aerosols in the gases leaving the combustion chamber, i.e. the fly ash. In the low-oxygen area of flames, NO can be formed via fuel radicals with nitrogen – so called "prompt NO_x".

Indoor nitrogen oxides air pollution can be generated by outdoor or indoor sources. The primary indoor sources of nitrogen oxides are unvented fuel burning appliances (gas ranges, kerosene space heaters etc.), heating appliances and tobacco smoking. Increased levels of nitrogen oxides can occur in homes with gas cookers, paraffi heaters and certain portable bottled gas heaters as well as with badly maintained central heating boilers or gas fires The rate of production is governed by three main factors: the amount of oxygen, the flame temperature, and the rate of cooling of the combustion products (since NO formation and conversion proceed more rapidly at higher temperatures). The average level in homes without combustion appliances is about half of that outdoor; in homes with kerosene heaters or unvented gas space heaters, indoor levels often exceed outdoor levels. Immediate concentrations of NO_x, occurring during cooking are still more dangerous, however, mainly during winter, when ventilation is insufficient.

Measurement of Nitrogen Oxides

Measurements of nitrogen oxides were carried out in order to determine indoor and outdoor NO_x concentrations during the year, as well as the influenc of air exchange and air tightness of building structures on indoor NO_x occurrence depending on the indoor sources of NO_x. The significant part of measurements of NO_x concentrations focused on the dynamics of NO_x concentrations changes during the combustion process for various rates of gas burning in modeled conditions for the purpose of determining relevant parameters influencing NO_x occurrence. The outcome of the modeled measurements was the development of a mathematical model for predicting the occurrence of indoor nitrogen oxides.

Method

The investigation included online gas measurement for NO, NO_2 and NO_x by the 'Thermo Environmental Instruments Inc. - model 42 - Chemiluminiscence NO-NO_2-NO_x Analyzer'. This analyzer was located in the middle of the room, at the level of the breathing zone of a sitting person - 105 cm above the floor. The results showed concentration peaks during cooking, but with large differences resulting from different rates of gas burnin

The principal method used for measuring NO_2 is based on chemiluminescence:

$$NO + O_3 \rightarrow NO_2{}^* + O_2$$

$$NO_2{}^* \rightarrow NO_2 + h\upsilon$$

Where, hυ represents a photon of solar radiation of energy and $NO_2{}^*$ is the NO_2 molecule in the excited state.

NO_2 concentrations are determined indirectly from the difference between the NO and NO_x (NO + NO_2) concentrations in the atmosphere. These concentrations are determined by measuring the light emitted from the chemiluminescent reaction of NO with O_3 (similar to the reaction of O_3 with ethylene noted for the measurement of O_3), except that O_3 is supplied at a high constant concentration, and the light output is proportional to the concentration of NO present in the ambient air stream.

Influence of Air Exchange on NO_x Concentrations

The effect of air exchange by ventilation on indoor nitrogen dioxide occurrence was studied in buildings with natural and forced air replacement during and after the combustion process. Monitoring of NO_2 concentrations was carried out during the use of a gas stove with gas consumption from 0.2 to 0.4 m³ for 30 minutes. The increase and decrease of NO_2 concentrations were monitored depending on the air exchange in the room. The measurements were performed with exhaust ventilation turned both off and on and at different adjustments of air exhaust. The maximum concentration of NO_2 with natural air exchange was 321.1 μg/m³, and 185.9 μg/m³ with an air exhaust rate of 130 m³/h, i.e. the air exchange rate was about 4 times per hour. Before the start of the combustion process, background concentrations of NO_2 ranging from 26.6 to 38.3 μg/m³ were determined, as were concentrations of NO_2 at the natural ventilation of 87.8 μg/m³ for 35 minutes after burning. In contrast, the value of NO_2 concentration reached only 20.3 μg/m³ with forced air replacement. The dynamics of the increase and decrease of NO_2 concentrations monitored in the room with natural and forced ventilation is evident from Fig.

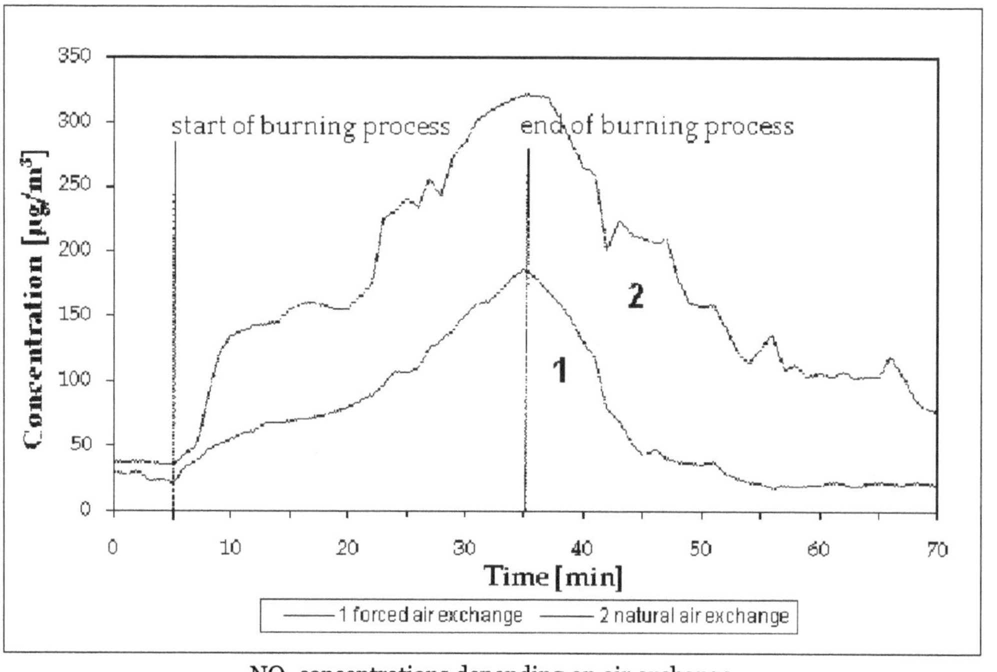

NO_2 concentrations depending on air exchange.

In rooms with a similar layout and volume, higher values of NO_2 concentrations were reached, especially when the combustion process did not provide sufficien air exchange (insufficien natural ventilation by infiltration). NO_2 concentrations exceeded the Slovak safety limit value of 200 μg/m³ during and after the combustion process. The measurements of NO_2 concentrations have confirmed that the level of nitrogen oxides indoors depends not only on the rate of combustion and on the volume of the room but also on the manner and rate of air exchange, and in particular on the specific distribution conditions

Influence of Building Structures on NO_x Concentrations

In order to verify the influence of air tightness of openings, NO_2 concentrations were measured in kitchens with both wooden and plastic windows. The gap permeability coefficien for wooden windows was i_{lv} = 1.4x10⁻⁴ m²/(s.Pa^{0,67}), the determined air exchange, n = 1.8. The gap permeability coefficien for plastic windows as determined by the manufacturer was i_{lv} = 0.02x10⁻⁴ - 0.4x10⁻⁴ m²/(s.Pa^{0,67}), the determined air exchange, n = 0.2. Measurements of NO_2 concentrations in buildings with wooden and plastic windows were made with other parameters (rates of gas burning, time and burning conditions etc.) equal. Outdoor NO_2 concentrations, depending on the time of measurement and diffusion conditions, varied from 17.3 μg/m³ to 36.1 μg/m³. The increase of NO_2 concentrations was monitored, depending on the intensity of gas burning, for sixty minutes and the decrease of NO_2 concentrations for ninety minutes. The influence of tightness of windows on the dynamics of NO_2 concentration changes for the intensity of gas combustion from 0.1 to 0.2 m³/h is shown in Fig.

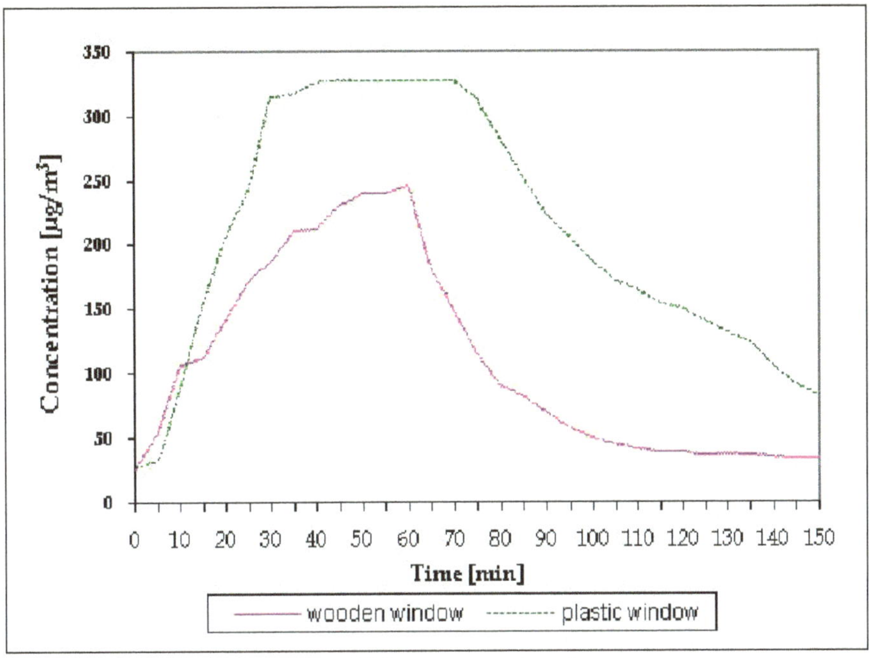

Dynamics of NO_2 concentration changes.

In kitchens with wooden windows a maximum NO_2 concentration of 246.1 μg/m³ was reached after 60 minutes of gas burning at a consumption rate of 0.1 to 0.2 m³/h. In kitchens with plastic windows, a significantly higher concentration of 315.2 μg/m³ was reached after 30 minutes of gas burning at a consumption rate of 0.1 to 0.2 m³/h. During the combustion process, the NO_2 concentrations increased linearly in both kitchens with wooden and plastic windows. After completion of the combustion process, the concentrations decreased and after 90 minutes a value of 33.7 μg/m³ was reached in kitchens with wooden windows. In kitchens with plastic windows the concentrations increased linearly for only the first 30 minutes, at which point they reached a value of 315.2 μg/m³. Then, for 45 minutes the concentrations steadied and values ranged from 318.7 to 328.4 μg/m³. Concentrations decreased 15 minutes after the end of the combustion process and a value of 82.4 μg/m³ was reached after 90 minutes. The dynamics of the changes in concentrations showed a need for a targeted air exchange during gas stove combustion.

Indoor NO_x Concentrations in Modeled Conditions

Measurements were made to detail NO_x production and determine the resultant indoor concentrations while gas appliances were being used. The intensity of gas combustion varied from 0.1 m³.h⁻¹ to 0.7 m³.h⁻¹. As kitchens had been identified as the most polluted rooms, measurements were taken there. The levels of background concentrations of NO, NO_2 and NO_x were monitored. The increase and decrease of NO_x concentrations was found at the same time.

Changes in NO_x concentrations for the range of gas consumption intensities were studied. The changes while allowing the combustion process to continue at a maximum rate for 60 min, and for decreasing rates of gas usage, were then followed for 90 min after combustion had stopped. The mean NO_x concentrations for the intensity of gas consumption from 0.1 m³.h⁻¹ to 0.7 m³.h⁻¹.

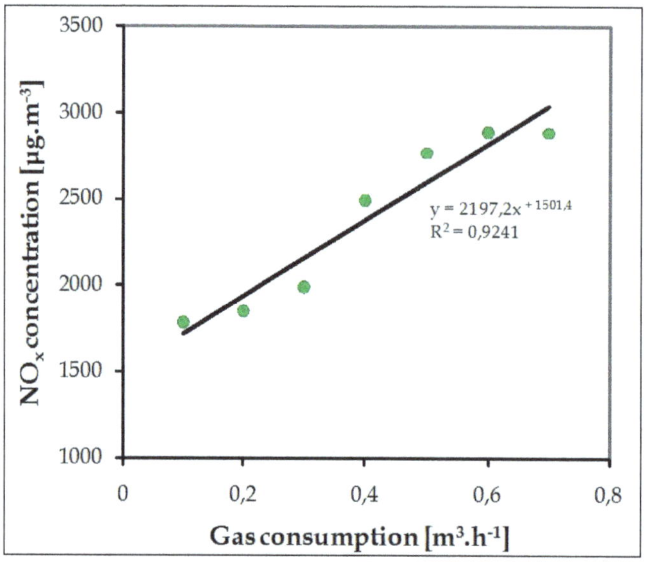

NO_x concentrations for the gas consumption range studied.

Before the start of the combustion process the average outdoor i.e. indoor background concentrations of NO_x were determined and mean values from 21.0 to 27.9 µg/m³ were recorded. The combustion process took 60 minutes and a linear increase was recorded for 5 - 60 minutes depending on the intensity of gas burning. After the completion of combustion, mean NO_x concentrations achieved high levels from 1785.9 to 2889.2 µg/m³. 90 minutes after the end of the combustion process, the mean values of NO_x were from 279.1 to 1293.6 µg/m³. The concentrations of NO_x after 1 h of gas combustion were shown to depend on the level of gas consumption ranging from 0.1 m³.h⁻¹ to 0.7 m³.h⁻¹.

The oxides formation at a gas consumption rate of 0.1 m³.h⁻¹ is given in Fig. It is evident that an essentially linear increase of NO, NO_2 and NO_x concentrations occurred during the whole combustion period to a fi al maximum value after 1 h, when gas usage was stopped (NO 1357.4 µg.m⁻³, NO_2 428.5 µg.m⁻³ and NO_x 1785.9 µg.m⁻³). A decrease in concentrations was then observed for a period of 90 min, after which the concentrations were NO 209.8 µg.m⁻³, NO_2 69.4 µg.m⁻³ and NO_x 279.1 µg.m⁻³.

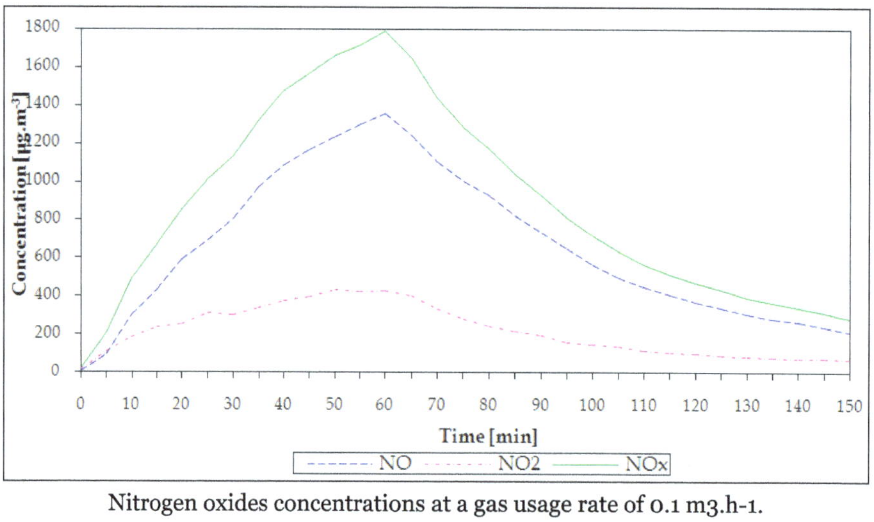

Nitrogen oxides concentrations at a gas usage rate of 0.1 m3.h-1.

Importantly, it should be noted that the concentration of NO_2 after 1 h of gas combustion (428.5 μg.m^{-3}) was more than double the Slovak safety limit (200 μg.m^{-3}). The limit was exceeded 10 min after the start of combustion (270.9 μg.m^{-3}) and concentrations remained above the limit for the next 100 min (228.6 μg.m^{-3}).

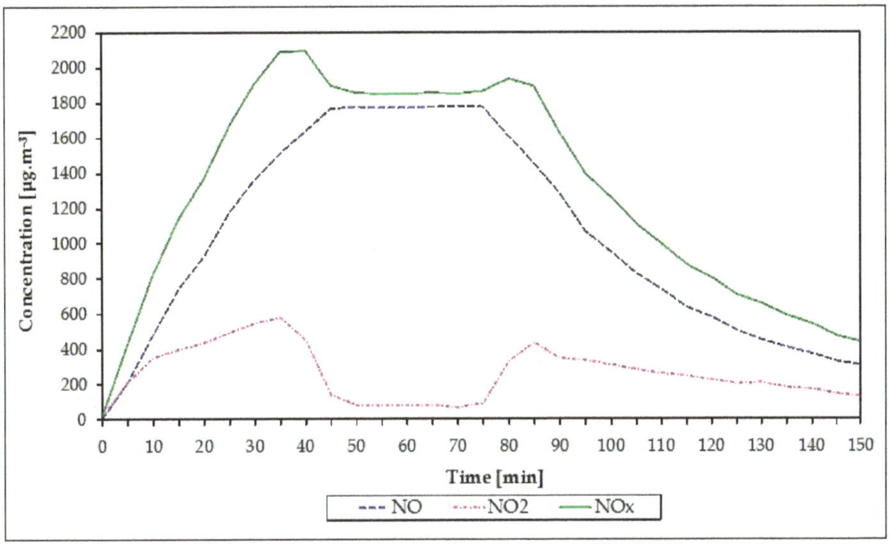

Nitrogen oxides concentrations at a gas usage rate of 0.2 m3.h-1.

The figure shows that a linear increase in concentrations of NO, NO_2 and NO_x was not sustained throughout the combustion process. Concentrations of NO_2 and NO_x rose linearly for 35 minutes, reaching an NO_2 value of 577.2 μg.m^{-3} and an NO_x value of 2089.8 μg.m^{-3}. Subsequently the concentrations decreased and for 25 minutes were almost constant, with NO_2 values ranging from 68.8 to 88.2 μg.m^{-3} and NO_x from 1851.0 to 1864.2 μg.m^{-3}. As a consequence of natural air exchange, an increase of NO_2 concentrations and decrease of NO concentrations occurred 15 minutes after the completion of combustion. 90 minutes after the end of combustion, the concentrations were NO 306.8 μg.m^{-3}, NO_2 134.8 μg.m^{-3}and NO_x 441.6 μg.m^{-3}.

Mathematical Modeling of Indoor Nitrogen Oxides

Modelling is a powerful tool that is being used extensively by scientists to investigate air movement and airflow patterns and to evaluate the performance of HVAC systems. Numerous models have been developed to study indoor air quality (IAQ). Although designers, architects, code enforcement officials and other building scientists have been using modelling as a powerful tool to address indoor IAQ issues, more development is needed to increase the accuracy and the validity of the IAQ models. These are being used to predict the dispersion of air contaminants within the building environment. Air pollution concentration can be estimated both spatially and as a function of time. The aim of IAQ models is to provide a way to link information about sources, distribution, dynamic changes of pollutant concentrations and also building factors to estimate indoor pollutant concentrations. Many of the latest indoor air models include exposure modelling and risk modelling. The type of model used depends on the planned uses of the model. The most common uses of IAQ models are the following: (i) estimating population exposure to various indoor pollutants, (ii) estimating the impact of individual sources on pollutant concentration and (iii) estimating the impact of individual sources and IAQ control options on personal exposure.

Designing a model and modelling can substantially contribute to problem solving and to obtaining new understanding of system behaviour with respect to air component dynamic changes. A mathematical description of the processes can generalise experimental results so that they may be used to interpret other processes physically similar to those introduced in the model. Although modelling doesn't achieve all the rules of similarity, it does provide overall information about a system's behaviour, which is only a little different from a real situation. Also, although it is only an approximation, the model does simplify description of the real physical measurements.

Most research institutes and universities focus on modelling and predicting outdoor and indoor pollutants. Most of the models developed to predict human exposures use empirical data on indoor/outdoor ratios (I/O) to determine indoor microenvironment concentrations as a function of time. These models are described in various studies. For example, one study describes a simple deterministic model, relating exposure to background ambient levels, indoor values, and human activities. Results suggest that indoor nitrogen dioxide concentrations in private dwellings vary primarily because of outdoor levels and the type of cooking fuel used, but are also affected by factors such as air-exchange rates and strength of indoor sources. Estimates of population exposures are obtained by combining observed distributions of nitrogen dioxide concentrations from outdoor and indoor settings with information about the number of people and amount of time spent in each microenvironment. One group of researchers developed regression models for indoor NO_2 exposure. Using analytical methods to choose descriptive proxies and evaluate the tradeoffs in their implementation can help epidemiological studies improve their designs. Other researchers developed and applied mathematical models to predict indoor air concentrations resulting from the transport of outdoor gas-phase nitrogen oxides, ozone and nitric acid indoors; and integral expressions that may be used to estimate model parameters from dynamic measurements of contaminant concentration. Another study deals with a modelling of indoor exposure to nitrogen dioxide in the UK. A dynamic multi-compartment computer model was developed to describe the physical processes determining indoor pollutant concentrations as a function of outdoor concentrations, indoor emission rates and building characteristics.

The model has been parameterised for typical UK homes and workplaces and linked to a time-activity model to calculate exposures for a representative homemaker, schoolchild and offic worker, with respect to NO_2. This model can be used to distinguish three types of exposures: outdoor exposure, indoor exposure resulting from penetration of outdoor air and indoor exposure resulting from indoor sources. The model predicts the above types of personal exposures for a representative homemaker, schoolchild and office worker, based on indicative activity pattern

The deterministic modelling approach applied here combines two types of models: (i) A physical model, used to calculate hourly indoor air pollutant concentrations for different microenvironments as a function of outdoor concentrations, building characteristics and indoor source emissions. This physical model, entitled INTAIR, is a simple dynamic compartment model which solves the resulting set of differential equations using a fourth-order Runge-Kutta scheme. (ii) An exposure model, used to calculate personal exposures by combining the movement of typical individuals through a series of microenvironments with the modelled ME concentrations.

Another study seeks to utilize publicly available data and questionnaire responses to predict residential indoor concentrations of traffic-relate air pollutants for lower socioeconomic status (SES) urban households. Indoor and outdoor 3–4 day samples of NO_2 were collected in 43 low SES residences across multiple seasons from 2003 to 2005. Home characteristics and occupant behaviours were collected via a standardized questionnaire. Additional housing information was collected through property tax records, and ambient concentrations were collected from a centrally located ambient monitor. The contributions of ambient concentrations, local traffi and indoor sources to indoor concentrations were quantified using regression analyses. Comparing models based on covariate selection using p-values or a Bayesian approach yielded similar results, with traffi density within a 50 m buffer of a home and distance from a truck route as important contributors to indoor levels of NO_2. The Bayesian approach also highlighted the uncertainty in the models.

The difficul of predicting NO_x emissions during combustion of fossil fuels is caused mostly by the complexity of the mechanisms of NO_x formation, mechanisms which depend on many factors. Some partial mechanisms have not yet been completely clarified so it is not possible to describe them exactly. Consequently, empirical correlations between fuel properties, combustion regime and NO_x emissions cannot be properly calculated. This affects the degree to which calculations can be informative in a mathematical model of indoor NO_x production. The final result has to be assembled from the results obtained from statistically suitable collections of measurements as well as more controlled laboratory measurements.

Input Parameters

The research task to describe NO_x production in the combustion process relies on the measurement of NO_x as influence by those parameters which directly characterize the combustion process, i.e. intensity of gas combustion I ($m^3.s^{-1}$), exposure period t (s), amount of indoor air m (kg) and air change n (s^{-1}).

All the given variables are defined using basic dimensions, allowing their direct use for dimensional analysis to create a mathematical model of NO_x production. All the variables can be simply measured during combustion allowing us to then compare NO_x production under the particular working conditions of the chosen combustion device on the basis of direct measurements and by means of the created mathematical model.

Dimensional Analysis Application

Dimensional analysis offers a method for reducing complex physical problems to the simplest form prior to obtaining a quantitative answer. At the heart of dimensional analysis is the concept of similarity. In physical terms, similarity refers to some equivalence between two things or phenomena that are actually different. Mathematically, similarity refers to a transformation of variables that leads to a reduction in the number of independent variables that specify a problem. The behavior of a physical system is defined by the complete set of dimensionless variables formed by the relevant physical variables. This fact suggests that if two systems have the same numerical values for all the defining dimensionless variables, then these two systems are dimensionally similar. Further, if there are two dimensionally similar systems, then their behavior can be closely correlated and hence the results of measurements on either one can be "projected" to the other. This is the basis for dimensional modelling. Therefore the first step in a model experiment is to construct a complete set of dimensionless variables relevant to the system. These dimensionless variables are then made to be equal for the model and the prototype.

The mathematical model describing NO_{xi} formation is based on the formation of dimensionless arguments π_i, from the stated variables that influenc formation of the oxides of nitrogen. The valuable property of these arguments is that in all existing systems of units they have the same numerical size and they have no dimension. The formation of a mathematical model rests on derivation of functional dependence from the expressed dimensionless variables, which in general always has an exponential character. Transformation of this function into logarithmic co-ordinates corresponds to a linear character that makes the work with the model easier and enables one to determine simply the parameters of linear function.

The model given here has universal validity for all combustion devices that have at least approximately similar geometric characteristics. For every combustion device the parameters of linear function, i.e. regression coefficients have to be determined separately. The general relationship between the selected variables which affect N_{xi} production can be expressed in the form:

$$\varphi(\mathrm{I}, \mathrm{t}, \mathrm{m}, \mathrm{NO_{xi}}, \mathrm{n}) = 0$$

The dimensional matrix-relationship created below has the rank of a matrix where r = 3 and its lines are dimensionally independent of each other. From n = 5 independent variables a matrix r can be set up with (n-r) dimensionless arguments.

$$\begin{array}{c|ccccc|} & t & I & m & NOxi & n \\ s & 1 & -1 & 0 & 0 & -1 \\ m & 0 & 3 & 0 & -3 & 0 \\ kg & 0 & 0 & 1 & 1 & 0 \end{array}$$

For the general form of the argument π is valid:

$$\pi = t^{x1}.I^{x2}.m^{x3}.NO_{xi}^{x4}.n^{x5}$$

From the condition that the left side of the identity equals one (ordinary number π), the resulting exponent of every basic dimension is equal to zero. By applying this condition for every element from r basic dimensions, we obtain a system of three linear equations with five unknowns. To solve the system, we have to select two unknowns (and this always twice) and then calculate the rest.

For the selected $x_4 = 1$, $x_5 = 0$ the exponents will be calculated from the linear equations system, thus $x_1 = 1$, $x_2 = 1$, $x_3 = -1$.

For the selected $x_4 = 0$, $x_5 = 1$ the exponents will be calculated from the linear equations system, thus $x_1 = 1$, $x_2 = 0$, $x_3 = 0$.

For these the dimensionless argument is then valid:

$$\begin{array}{c|ccccc|} & x1 & x2 & x3 & x4 & x5 \\ \pi1 & 1 & 1 & -1 & 1 & 0 \\ \pi2 & 1 & 0 & 0 & 0 & 1 \end{array}$$

The required dimensional homogeneous function in dimensionless form is:

$$\Psi(\pi 1, \pi 2) = 0$$

or after adjustment and backward transformation of dimensions for the particular variables, the function will have the form:

$$\Psi\left(\frac{t.I.NOxi}{m}, t.n\right) = 0$$

The real course of the dependence of dimensionless arguments π_1 to π_2 calculated according to the relationship from the measured values:

$$\pi_1 = A.\pi_2^B$$

From this a regression straight line for the calculation of the regression coefficient can be calculated by the method of least squares. After completing relationship using relationship and its modification, we obtain the relationship characterizing N_x production in the form:

$$NOxi = \frac{10^9.A.t^{B-1}.n^B.m}{I}$$

Where, A and B are regression coefficients calculated by the method of least square

The mathematical formula presents a mathematical model of NO_{xi} production through using of gas appliances in the kitchen. The indoor NO_x is given by the formula:

$$NO_x = NO_{xi} + NO_{xo}$$

NO_{xi} – Indoor nitrogen oxides concentration [$\mu g.m^{-3}$].

NO_{xo} – Outdoor nitrogen oxides concentration [$\mu g.m^{-3}$].

Verification of the Mathematical Model

The mathematical model was verified for various amounts of gas consumption in the range of 0.1 – 0.7 $m^3.h^{-1}$. As an illustration, the verification for gas consumptions of 0.1 $m^3.h^{-1}$ and of 0.2 $m^3.h^{-1}$ using a model experimental room is given below. The amount of air was 28.28 kg and the air change rate 3.56x10-4 s^{-1}. The air change rate was determined on the basis of decreasing NO_x concentration per hour.

On the basis of measured input parameters, NO_{xi} concentrations were determined through the located constants and regression coefficient obtained by the method of least squares. These values are presented in Table.

Table: Regression coefficient value

Gas consumption [$m^3.h^{-1}$]	A	B
0.1	4.0000E	1.8420
0.2	2.0000E	1.7765

0.3	2.0000E	1.8179
0.4	3.0000E	1.8593
0.5	4.0000E	1.6603

As an illustration, the relationship for NO_{xi} production at a gas consumption rate of 0.1 m³.h⁻¹ is obtained from:

$$NOxi = \frac{4.0E-09.t^{1.8420-1}.n^{1.8420}.m}{I}$$

The values of measured and calculated NO_{xi} concentrations are compared in Table. The differences between measured and calculated NO_{xi} concentrations are considered acceptable. The deviations reflec two things: the selection of relevant parameters is a simplification; their degree of involvement and the number of factors on which NO_{xi} production depends, which they represent, are both unknown.

The relationship for NO_{xi} production at a gas consumption rate of 0.2 m³.h⁻¹ is obtained from:

$$NOxi = \frac{2.0E-09.t^{1.7765-1}.n^{1.7765}.m}{I}$$

A linear increase of NO_{xi} concentrations continued for forty minutes from the start of combustion, i.e. only this amount of time was required for complete combustion of gas burning.

Table: Measured and calculated NO_{xi} concentrations for gas consumption of 0.1 m³.h⁻¹.

t [min]	$NO_{xi, measured}$ [μg.m⁻³]			NOxi, calculated [μg.m⁻³]	Deviation [%]
	min	max	mean		
5	195.7	219.5	204.7	243.9	20.67
10	426.7	534.5	489.5	420.6	16.34
15	593.4	740.3	671.5	583.2	16.97
20	712.7	992.5	854.5	737.3	19.54
25	884.5	1161.7	1013.0	885.4	18.43
30	999.8	1262.0	1131.8	1028.9	17.1
35	1213.4	1478.5	1320.8	1168.6	13.89
40	1410.5	1566.1	1477.9	1305.1	12.15
45	1499.4	1645.3	1565.1	1439.0	10.77
50	1571.1	1747.4	1660.9	1570.5	10.02
55	1639.6	1776.3	1711.7	1700.0	11.53
60	1762.1	1803.3	1785.9	1827.7	11.25

Table: Measured and calculated NO_{xi} concentrations for gas consumption of 0.2 m³.h⁻¹.

t [min]	$NO_{xi,measured}$ [μg.m⁻³]			NOxi,calculated [μg.m⁻³]	Deviation [%]
	min	max	mean		
5	316.2	556.0	429.2	421.2	33.89
10	632.6	1016.2	829.8	705.6	24.98

15	1003.8	1290.2	1140.4	958.5	15.16
20	1230.8	1499.1	1366.1	1192.8	13.26
25	1527.1	1779.5	1669.2	1414.3	14.80
30	1785.4	2031.0	1910.1	1625.9	14.54
35	1960.2	2201.3	2089.8	1829.9	12.00
40	2048.6	2151.4	2092.7	2027.3	4.48

Interpretation of Mathematical Model

Current knowledge as well as the results of monitoring in a selected group of residential buildings in situ confir ed the significant occurrence of indoor nitrogen oxides, especially with the burning of gas in kitchens, and lead to the construction of a mathematical model for predicting the occurrence of nitrogen oxides. Subsequently, a mathematical model based on the dimensional analysis method was verified for each intensity of gas burning. The calculated values of NO_{xi} concentrations were compared with average measured concentrations of NO_{xi}. This difference was established as an absolute error which ranged from values of 4.48% - 34.05%. The average absolute error is 17.86%. This value is acceptable, and confirms that the proposed mathematical model is useful for predicting the occurrence of indoor nitrogen oxides.

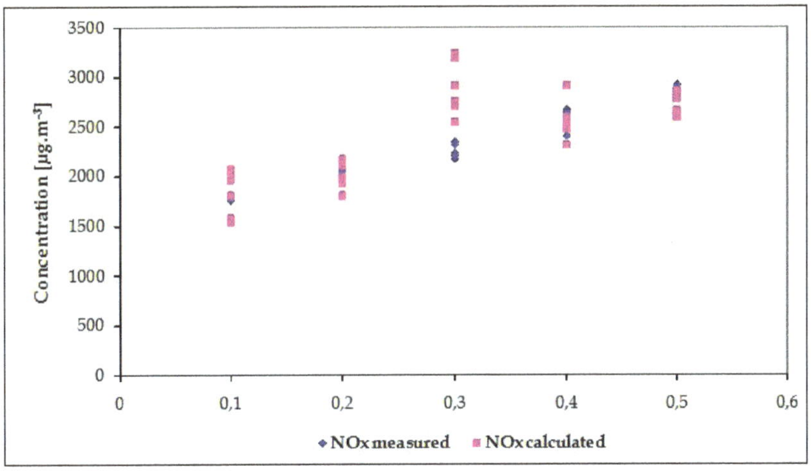

Variance of measured and calculated NO_{xi} concentrations for various gas intensities.

Urban Air Pollution

Air pollution represents a health risk for people living in urban environment. Urban air consists in a complex mixture of chemicals and carcinogens and its effects on health can be summarized in acute respiratory effect , neoplastic nonneoplastic (e.g. chronic bronchitis) chronic respiratory effects, and effects on other organs and system

There has been a massive growth in the size and the population of the cities, in particular, in developing countries. An endless stream of automobiles is being added each year. This has led to deterioration in the quality of air in the cities and neighboring areas. The air pollution has increased. The urban air pollution is an important factor in changing the climate of cities.

The following urban characteristics seriously affect the urban climate

- Creation of urban heat islands.
- Lower average wind speeds.
- Increase in fog frequency.
- Changes in the vertical structure of the lower boundary layer.
- Formation of meso-scale circulation systems, etc.

The major pollution factors, which are briefly discussed below, are waste heat, sulfur dioxide, carbon monoxide, nitrogen oxides, hydrocarbons/organic compounds, photochemical oxidants and suspended particulate matter.

Waste Heat

Waste heat is an important aspect of urban temperature and moisture content. It is unique to urban environments due to dense network of sources of heat. It is likely to aid in the formation of secondary pollutants.

Sources of Waste Heat

- Automobiles and transportation exhausts.
- Industrial emissions.
- Power plant plumes.
- Domestic heating/cooking.
- Air conditioning.
- Excess heat stored in buildings, pavements, roads and other structures which heat up fast.

In urban centers due to lack of plant cover only small fraction of this heat is used for evapo-transpiration. Most of the heat is released as sensible heat and small portion as infrared radiation and latent heat.

The waste heat flux is determined largely by density of sources. The average heat released by cities shows the large spatial variation. For example, the annual heat released by Manhattan city has been reported to be about 17.4 GW (in summer) and 21.8 (in winter). The heat fluxe varied widely from 5 W/m^2 on the urban fringes to >100 W/m^2 in the city centre. These values are large when compared with annual average net radiation balance at earth surface in mid - latitude of about 100 W/m^2. Thus, the cities add a significant portion of artificial heat to radiation balance.

Obviously, the temperature in the city is likely to be higher than the surrounding areas. However, the total contribution of cities to earth radiation balance on global scale is negligible. Nevertheless, at the local scale this is serious.

Automobile Exhaust Emissions

The major pollutants released by automobile exhaust are as follows:

- Carbon monoxide, CO: It is the most abundant pollutant in urban areas. It is formed due to incomplete combustion of carbonaceous fuels. The correct air to fuel ratio is ~ 14.5. If the mixture is rich (more fuel), there is insufficien oxygen to completely oxidize carbon into CO_2 resulting in the formation and emission of CO.
- Unburnt hydrocarbons: In urban areas the exhaust emissions, evaporative losses from fuel tank, carburetor etc. are the major sources of polycyclic aromatic hydrocarbons (PAHs) and volatile organic compounds (VOCs). Many of these are highly toxic and carcinogenic. Ethylene is produced in petrol engines due to use of ethylene dibromide and dichloride.
- Sulfur Dioxide, SO_2: This is formed due to oxidation of sulfur present in fuel.
- Nitrogen Oxides, NO_x: Due to high temperature inside internal combustion engine, N_2 and O_2 combine to form NO_x.
- Organic / Inorganic Lead – Compounds: Use of tetraethyllead(TEL)(now banned) as an-ti-knock in petrol was responsible for their release in atmosphere.
- Smoke/Soot: Particularly in diesel vehicles, the emission of the soot /smoke is most visible. It is formed due to incomplete oxidation of carbon. The soot particles are polymeric substances having carbons chains (cyclic). The soot particle have free valance on surface and so these have great sticking property.
- Particulates.
- Odor: This is due to partial oxidation of hydrocarbons into aldehydes. Petrol engines release larger amounts of CO (~3%) and unburnt HC. Diesel engines release more of NO_X, smoke and odor.

Monitoring of Carbon Monoxide and Smoke

Monitoring of CO is done through non -dispersive infrared (NDIR) method. Infrared radiation are generated through a hot filament, and then passed through a sample and a reference alternately. Hatridge smoke meter, which is based on light extinction method, is used. The sample and reference are passed alternatively through a tube, at the one end of which is a light source and at the other end a photocell. The light reaching a photocell is converted into electric current, which is measured.

Control Measures

- Proper maintenance of engine.
- Petrol tank should be full. There are greater evaporative losses when the tank is only partially full.
- Use of catalytic converter in vehicles.

Air Pollution Monitoring using Fuzzy Logic

Monitoring by fuzzy logic means air pollutant risk will be evaluated on time in the specific industry and area. It will be able to identify which air pollutant has high risk and which industry is responsible for that, and also, which area is in the matter. In other word, it will evaluate air pollutant risk, related industry and related area on time. Therefore, it helps Air Pollution Control District (APCD) to identify high risk air pollutant, industry and area by numerical value. Fuzzy logic is used in uncertainty environments so that expert's experiments can be converted into mathematical languages. On base of fuzzy logic is possible to maximize the utilization of design.

Fuzzy logic has a real concept of natural actuation which everybody meets every day. This logic seems to be natural phenomena in a real form. Fuzzy logic is used in uncertainty environments so that expert's experiments can be converted into mathematical languages. On the bases of fuzzy logic it is possible to maximize the utilization of decision making. It is possible to use it also for ranking of air pollutant risk, dangerous industries and high risk area by a continuous value between 0 and 1.

Fuzzy logic investigates the relative air pollutant. In order to accomplish this, fuzzy-logic approach defines a set for each property. For example, air pollutant has different criteria; industries produce different amount of air pollutant and also the mass of industry production differences in the specific area. Therefore, these matters have different membership degree in their sets. Using these sets and fuzzy logic rules, it can evaluate and compare different air pollutant risk for APCD decision making properly.

Fuzzy logic method was proposed because this method has not been used for air pollutant risk ranking. In the other word, fuzzy logic method shows the satisfactory value of air pollutant in a continuous value between 0 and 1.

In the first step, the parameters must be normalized. The second step is sets definition. These are 3 sets called air pollutant set, industry set and area set. In the final step, the air pollutant favorability degree in the specific industry at the relative area is determined; the industry favorability degree for the specific air pollutant production at the relative area is determined and also the area favorability degree for the healthy environment in the specific industry at the relative air pollutant is determined.

Fuzzy logic explains the relativity of different features and defines a set for each feature and then determines the desirability of each variable to provide the involved feature on a scale of zero to one. This value, called the dependency degree of the involved variable in relation to the mentioned feature, will be evaluated.

Zero desirability degree means that the involved variable isn't provided at all for the related variable or the involved variable is not desirable for the correspondent feature at all; and desirability degree one implies that the target feature is completely provided for the involved variable or the involved variable is absolutely desirable for the related feature. A desirability degree of between zero and one means that the involved feature is partially provided for the target variable or the involved variable is partially desirable for the related feature.

There are different methods like AHP, Fuzzy Hierarchical TOPSIS and Fuzzy network TOPSIS for decision making. All these methods are heavily dependent on expert's opinion which does not have a scientific basis in general while fuzzy logic has a regulated pattern and is rarely influenced by expert's opinion.

The function that yields the accuracy or inaccuracy of propositions in fuzzy logic is called the membership function. The defined membership function should be compatible with the nature of process; otherwise, the fuzzy inference will be erroneous. It is thus necessary to identify the process completely before defining the membership function

How does fuzzy infer these figures then? There are different rules to infer fuzzy. One of the most famous ones is the fuzzy maximum-minimum rule (fuzzy peak-trough). It should be noted that the inference rule has to be compatible with the nature of process. Having sampled inferences (especially at limit quantities) and compared them with reality (real quantities), we can make sure of the correctness of selecting the fuzzy inference rule.

The minimum rule is used when a connector i.e. the conjunction 'and' is used. In this rule, the minimum value of propositions is accepted as an inference; in other words, if there are some propositions valued a, b, c, d,..., the minimum rule infers the minimum of these values.

Fuzzy Minimum Rule = min(a, b, c, d,...)

The maximum rule is used when a separator i.e. the conjunction 'but' is used. In this rule, the maximum value of propositions is accepted as an inference; in other words, if there are some propositions valued a, b, c, d, ..., the maximum rule infers the maximum of these values.

Fuzzy Maximum Rule=max(a, b, c, d,...)

Comparison between Binary Logic and Fuzzy Logic

Einstein says "where classic mathematics rules are realized, they are not reliable and where they are reliable, they cannot denote reality." Einstein's description is an imagination of the inefficienc of classical logic rules in mathematics. In natural conditions, indeed, the world of zero and one, correct and incorrect, good and bad is an abstract and imaginative world. A given subject is rarely 100 percent correct or incorrect and good or bad.

Using fuzzy logic, the precision of results increases and errors decrease. For instance, using the limited elements method, surface residual stresses resulting from grinding with an average error of 12.3 percent are predictable; however, the error level of calculation decreases to 0.037 percent by using this logic.

We are familiar with the term 'fallacy' in binary logic. Its famous example is that if the proposition 'the fire is on the ship' is a true proposition and if the phrase 'the ship is on the water' is true too, the proposition 'the fire is on the water' should be true as well. Fire and water are two opposite things in reality and cannot be together;. It is obvious that the final conclusion is wrong because the value of these propositions is correct or incorrect in our view and there is no an intermediate state and also we have not considered the opposite natures of water and fire

Fuzzy logic deals with this process as it is. It means that it gives a value between zero and one for the accuracy of the proposition 'the fire is on the ship'. Value one is given when fire has covered a ship completely and value zero is given when there is no fire on the ship. Generally, the degree of the accuracy of this proposition is defined according to the fraction of fire on the ship. There is a similar view for the proposition 'ship on water' i.e. a fraction of ship touching water is regarded as the degree of the accuracy of the proposition (the ship is on the water). Consequently, when the ship is completely sunk, this proposition is true with a value of one, but when the ship does not touch water at all, this proposition is true with a value of zero.

The opposite natures of water and fire are occult in fuzzy logic. Given what was said so far, if the fire has covered 30 percent of the ship, the proposition 'the fire is on the ship' is true with a value of 0.3. Also, if 20 percent of the ship is touching water, the proposition 'the ship is on water' is true with a value of 0.2. According to the fuzzy minimum rule, the proposition 'the fire is on the water' is then true with the least value of the propositions which is 0.2 percent.

$$\text{Min}\ (0.3, 0.2) = 0.2$$

Notice that the defined membership functions are compatible with the nature of processes. if water covers all the ship which means that the ship is sunk, the proposition 'the ship is on the water' would be true with a value of one. But certainly there would be no fire on the ship in this case since water and fire are opposites. The proposition 'the fire is on the ship' is then true with a value of zero and, therefore, the proposition 'the fire is on the water' is true with the least values which is zero.

$$\text{Min}\ (1, 0) = 0$$

Monitoring of Industrial Pollutants

Monitoring of pollutants can be studied in three sets. The first set is the type of the metallurgical process that provokes pollution, so set M, the set of metallurgical processes {iron and steel production, aluminum production, copper production, zinc production} is defined as relation 3. The second set is an area that provokes pollution; so a set of areas is formed like relation 4. Set A, is a set of areas definable all over the universe. For example, the set of world factories {Khuzestan Steel, Isfahan Iron Melt, Bohler, Myhanaite} or the set of world states or provinces {Mazandaran, Gilan, Florida, Texas} or the set of world countries {Iran, Japan}. Also, the set of areas can be defined in a country e.g. the set of factories in Iran {Khuzestan Steel, Isfahan Iron Melt} or the set of provinces in Iran {Mazandaran, Gilan}. More general the selected area (the whole world) and more detailed its divisions or the members of the set (factories) are, more detailed the monitoring will be. Due to the insufficienc of data on the area set in this research, the set of countries was taken into account. The third set is the type of pollutant that led to the formation of set P, the set of pollutants {carbon dioxide, carbon monoxide, sulfur dioxide, lead, nitrogen oxides, evaporative organic compounds, arsenic, lead and copper in water} according to equation number 5. Set M can have m members, set A can include a members and set P can have p members.

$$M = \{i\nabla N, i\nabla[1,m], m_i\} = \{m_i\}$$

$$A = \{j\nabla N, j\nabla[1,a], a_{i,j}\} = \{a_{i,j}\}$$

$$P = \{k\nabla N, k\nabla[1,p], P_{i,j,k}\} = \{P_{i,j,k}\}$$

m_i is the process i we want to study. $a_{i,j}$ is the process i in area j which will be studied and $P_{i,j,k}$ is the amount of pollutant k in process i that is in area j. It should be noted that the amount of pollutant is normalized i.e. $P_{i,j,k,\ Normalized}$ amount of pollutant is produced in area j for each unit of product of process i. Hence, equation 6 is used to calculate $P_{i,j,k,\ Normalized}$.

$$P_{i,j,k,Normalized} = \frac{\text{total k pollution in the } a_{i,j}}{\text{total mass production of} a_{i,j}}$$

Safety Degree Determination

The goal is to determine the biocompatibility or safety degree of process i in producing pollutant k in area j. Therefore, a new set named C is formed which means biocompatible processes in producing different pollutants in different areas. Each member in this set belongs to set C at a certain degree. This degree indicates the degree of biocompatibility or safety of a certain process in producing a certain pollutant in a certain area i.e. $C_{i,j,k}$. Equation below which is called the membership function of the members of the set C, i.e. $C_{i,j,k}$, is used to calculate $C_{i,j,k}$:

$$C = \{Ci, j, k\}$$

$$\text{If } C_{i,j,k} > 0 \text{ then } C_{i,j,k} = \left(\frac{\min\min\left\{P_{i,j,k,Normalized}\right\}}{P_{i,j,k,Normalized}} \times \left(1 - \frac{G_{k,j}}{G_{k,critical}}\right)\right) \text{Else } C_{i,j,k} = 0$$

$G_{k,j}$ is the amount of pollutant k in area j and $G_{k,critical}$ is the critical amount of pollutant k that is reported by the Air Pollution Control District (APCD) and there is thus a fixed and certain amount for each pollutant. $G_{k,j}$ is equal in equilibrium conditions in all over the world; however, this amount is different in real conditions for different area

Thus, all processes are scaled continuously between 0 and 1 relatively according to the pollutants they produce and in relation to the area which is located in. As the scale 0 for the member $C_{i,j,k}$, i.e. the process i located in area j, is not biocompatible or safe at all in producing pollutant k, while scale one for member $C_{i,j,k}$, i.e. process i located in area j, is completely biocompatible or safe in producing pollutant k and a scale between 0 and 1 for member $C_{i,j,k}$, i.e. process i located in area j, is partially biocompatible or safe in producing pollutant k.

The worst conditions are considered in this research; therefore, the minimum fuzzy logic rule is used. The most dangerous process i in producing pollutant k in area j is the process that has the least biocompatibility or safety in producing pollutant k in area j. As a result, $C_{i,j,k,criticalis}$ obtained from the fuzzy minimum rule like equation below:

$$C_{i,j,k,critical} = \min\left\{C_{i,j,k}\right\}$$

Perhaps, it seems that monitoring of pollutants only through the set of processes is confusing. For instance, areas having advanced technologies have higher degrees of biocompatibility or safety because the normalized amount of the pollutant is low; but the total amount of the production of process at a unit of time may be so high that it practically includes a major part of the pollutant production. It is noteworthy that if the amount of the product in areas with advanced technologies reduces, with which technology would the shortage caused by the reduction in production be produced? If the involved technology is not as advanced as the intended technology, the amount of the pollutant will increase more than the past in equilibrium conditions. Hence, the total product production amount variable in an area will be placed in the second statement of equation

$$\text{If } C_{i,j,k} > 0 \text{ then } C_{i,j,k} = \left(\frac{\min \min \left\{ P_{i,j,k,\text{Normalized}} \right\}}{P_{i,j,k,\text{Normalized}}} \times \left(1 - \frac{G_{k,j}}{G_{k,\text{critical}}} \right) \right) \text{Else } C_{i,j,k} = 0 \text{ over time.}$$

Air Pollutants

An air pollutant is known as a substance in the air that can cause harm to humans and the environment. Pollutants can be in the form of solid particles, liquid droplets, or gases. In addition, they may be natural or man-made. Pollutants can be classified as either primary or secondary. Usually, primary pollutants are substances directly emitted from a process, such as ash from a volcanic eruption, the carbon monoxide gas from a motor vehicle exhaust or sulphur dioxide released from factories.

Secondary pollutants are not emitted directly. Rather, they form in the air when primary pollutants react or interact. An important example of a secondary pollutant is ground level ozone — one of the many secondary pollutants that make up photochemical smog.

Classification of Pollutants

The pollutants can be classified in distinct groups in a variety of ways, some of which are Primary Pollutants and Secondary Pollutants.

Primary Pollutants

Primary pollutants cause pollution by their direct release into the environment . The substance released may already be present in some quantities, but it is considered a primary pollutant if the additional release brings the total quantity of the substance to pollution levels. For example, carbon dioxide is already naturally present in the atmosphere, but it becomes toxic when additional releases cause it to rise above its natural concentrations. The rise in levels of carbon dioxide is one contributing factor to the greenhouse effect

The direct release into the atmosphere of a chemical that is not normally present in the air is also classified as a primary pollutant. Hydrogen fluoride, for example, is released from some coal-fired furnaces, but the compound is not usually present in unpolluted air.

A primary pollutant can be generated by many sources: pesticide dust sprayed intensively in agricultural areas, emissions from car and industrial exhausts, or dust kicked about from mining operations, to name a few. Some primary pollutants are composed of particulate matter that is not easily dispersed. Smoke , soot, dust, and liquid droplets released into the air either by the burning of fuel or other industrial or agricultural processes, are considered primary pollutants.

Primary pollutants also originate from natural sources. Volcanic ash, as well as grit and dust from volcanic explosions belong to this category. Salt dust blown inland by strong oceanic winds or gaseous pollution that originates from bogs, marshes, and other decomposing matter can also be classified as primary pollutants

When two or more primary pollutants react in the atmosphere and cause additional atmospheric pollution, the result is called secondary pollution. Nitrogen oxides, for instance, can react with volatile organic compounds to from smog , a secondary pollutant.

These are released directly into the air from the source. Based on physical state, these are classified as follows:

Gaseous Primary Pollutants

Some of the gaseous pollutants are as follows:

- Sulfur dioxide, SO_2.
- Nitrogen oxides, NO_x. ($NO_x = NO + NO_2$).
- Hydrocarbons (CH_4, propane, ethylene, etc); volatile organic compounds (VOCs); polycyclic aromatic hydrocarbons(PAHs), and large number of other organic compounds.
- Carbon monoxide, CO.
- Tropospheric ozone, O_3.
- Carbon dioxide, CO_2, chlorofluorocarbons (CFCs), halons, nitrous oxide, N_2O. Although these pollutants as such do not have any detrimental effect but are implicated in global climate change.

Primary Aerosols

Aerosols are dispersions of solid /liquid particulates in air. The primary particulates like soil particles, fly ash, smoke etc. are directly released in air

Depending on the source/origin air pollutants are classified as

(a) Primary air pollutants and (b) Secondary air pollutants.

Primary Air Pollutants

Primary air pollutants are those that are emitted directly from the identifiable sources. The examples of primary air pollutants:

- Carbon monoxide.
- NO_x (oxides of nitrogen).

- SO_x (oxides of sulphur).
- Hydrocarbons.
- Particulate matter.

Secondary Pollutants

Secondary pollutants are not emitted directly to the air, water, or soil. Secondary pollutants are synthesized in the environment by chemical reactions involving primary, or emitted chemicals. The best known of the secondary pollutants are certain gases that are synthesized by photochemical reactions in the lower atmosphere. The primary emitted chemicals in these reactions are hydrocarbons and gaseous oxides of nitrogen such as nitric oxide and nitrogen dioxide. These emitted chemicals participate in a complex of ultraviolet-driven photochemical reactions on sunny days to synthesize some important secondary pollutants, most notably ozone, peroxy acetyl nitrate, hydrogen peroxide, and aldehydes. These secondary compounds, especially ozone, are the harmful ingredients of oxidizing or photochemical smogs that cause damages to people and vegetation exposed to this type of pollution.

Most ozone is found in the upper atmosphere, where it acts to screen out much of the harmful radiation from the sun. Upper-level ozone is an important part of Earth's life support system. Lower-level ozone is created when sunlight hits hydrocarbons and nitrogen oxides released into the lower atmosphere by industrial and natural processes. Ozone is well known as an irritant to human respiratory systems, as a strong oxidant that causes materials to age rapidly and degrade in strength, and as a toxic chemical to plants. In terms of causing damage to agricultural and wild plants, ozone is the most damaging air pollutant in North America. Low-level ozone also acts as a greenhouse gas, restricting the escape of heat from Earth's surface and thus contributing to the global warming process.

Scientists estimate that the amount of low-level ozone currently in Earth's atmosphere is 100-200 times higher than it was only 100 years ago. The formation of low-level ozone can be slowed by reducing emissions of human-created hydrocarbons and nitrogen oxides into the atmosphere. Reducing hydrocarbon emissions by using catalytic converters on vehicles and generally reducing automobile travel time helps, as does the use of filtering devices to scrub industrial air emissions. However, many filters and converters fail to remove nitrogen oxides from emissions, and human-produced nitrogen oxides can combine with naturally produced hydrocarbons just as easily as with human-produced hydrocarbons. Planting trees and plants doesn't help, but the development and installation of converters and filters for removing both hydrocarbons and nitrogen oxides can.

Secondary pollutants can also be formed in other ways. For example, when soils and surface waters become acidified through atmospheric depositions or other processes, naturally occurring aluminium in soil or sediment minerals becomes more soluble and therefore, becomes more available for uptake by organisms. The soluble, ionic forms of aluminium are the most important toxic factor to plants growing in acidic soils and to fish in acidic waters. In this context, aluminium can be considered to be a secondary pollutant because it is made biologically available as a consequence of acidification

A few pesticides generate toxic chemicals when they are chemically transformed in the environment, and this phenomenon can also be considered to represent a type of secondary pollution. For example, dithiocarbamate is a fungicide used in the cultivation of potatoes. Ethylene thiourea is an important metabolite of this chemical, formed when the original fungicide is broken down by microorganisms in soil. Ethylene thiourea is relatively stable in soils and also somewhat mobile so that it can leach into ground water. Ethylene thiourea has been demonstrated to be carcinogenic in mammals, and it therefore represents an important type of toxicity that was not characteristic of the original fungicide.

The pollutants that are produced by the interaction among two or more primary air pollutants in the air or by reaction with constituents of air. These pollutants are formed in air due to the photochemical and thermal reactions of primary pollutants. For example, aerosols, ground level ozone, peroxyacetyl nitrate (PAN), tropospheric ozone, acid rain, photochemical smog are the pollutants of this type.

Secondary Gaseous Pollutants

These are generated in air through chemical reactions. Some of these are as follows:

- Tropospheric ozone, O_3.
- Aldehydes.
- Peroxyacetylnitrate etc.

Secondary Aerosols

These are formed in air, mostly through gas to particulate conversion, e.g., SO_2 to sulfate aerosol, NO_x to nitrate aerosols, ammonia into ammonium sulfate aerosol, VOCs into carbonaceous particles etc.

Based on the origin of aerosols, these are classified as follows

- Biological Aerosols: These are of biological origin, viz., seeds, pollens, spores, fragments of plants/ animals/insects, hair, bacteria, fungi, micro-organisms, algae, protozoa, viruses etc.
- Geochemical Aerosols: These aerosols are made up of geo-chemicals and are directly injected into air due to volcanic eruptions, mining, crushing, blasting, dispersion of dust through wind or human activities.
- Oceanic Aerosols: Ocean is a big source of aerosols such as marine salt, NaCl, as aerosol.
- Anthropogenic Aerosols: Fossil fuel combustion, biomass burning, industrial activities, e.g., manufacturing and metallurgical processes, cement plants etc. are some of the big sources. Automobiles are other big sources of this type of aerosols.
- Other Sources of Aerosols: Some other sources of aerosols are forest fires and meteoritic debris.

Based on chemical composition the aerosols are classified as follows

- Inorganic Aerosols: Basically, these are made from inorganic compounds and further divided in two categories:
    - Soluble Aerosols: These are soluble in water. The aerosols of salts like NaCl, and $(NH_4)_2SO_4$ are of this type.
    - Insoluble Aerosols: The examples of aerosols this type includes those derived from minerals, stones, etc. These are not soluble in water.
- Organic Aerosols: These are made from organic compounds. Examples are carbonaceous particles, polycyclic hydrocarbons (PAHs), soot, etc.

Major Air Pollutants

Criteria Pollutants

Clean, dry air consists primarily of nitrogen and oxygen—78 percent and 21 percent respectively, by volume. The remaining 1 percent is a mixture of other gases, mostly argon (0.9 percent), along with trace (very small) amounts of carbon dioxide, methane, hydrogen, helium, and more. Water vapour is also a normal, though quite variable, component of the atmosphere, normally ranging from 0.01 to 4 percent by volume; under very humid conditions the moisture content of air may be as high as 5 percent.

There are six major air pollutants that have been designated by the U.S. Environmental Protection Agency (EPA) as "criteria" pollutants—criteria meaning that the concentrations of these pollutants in the atmosphere are useful as indicators of overall air quality. The sources, acceptable concentrations, and effects of the criteria pollutants are summarized in the table

Criteria air pollutants				
Pollutant	Common sources	Maximum acceptable concentration in the atmosphere	Environmental risks	Human health risks
Carbon monoxide (co)	Automobile emissions, fires, industrial processes.	35 Ppm (1-hour period); 9 ppm (8-hour period).	Contributes to smog formation.	Exacerbates symptoms of heart disease, such as chest pain; may cause vision problems and reduce physical and mental capabilities in healthy people.
Nitrogen oxides (no and no_2)	Automobile emissions, electricity generation, industrial processes.	0.053 Ppm (1-year period).	Damage to foliage; contributes to smog formation.	Inflammation and irritation of breathing passages.
Sulfur dioxide (so_2)	Electricity generation, fossil-fuel combustion, industrial processes, automobile emissions.	0.03 Ppm (1-year period); 0.14 Ppm (24-hour period).	Major cause of haze; contributes to acid rain formation, which subsequently damages foliage, buildings, and monuments; reacts to form particulate matter.	Breathing difficulties, p - ticularly for people with asthma and heart disease.

Ozone (o_3)	Nitrogen oxides (no_x) and volatile organic compounds (vocs) from industrial and automobile emissions, gasoline vapours, chemical solvents, and electri-cal utilities.	0.075 Ppm (8-hour period).	Interferes with the ability of certain plants to respire, leading to increased susceptibility to other environmental stressors (e.G., Disease, harsh weather).	Reduced lung function; irritation and inflammation of breathing passages.
Particulate matter	Sources of primary particles include fires, smokestacks, construction sites, and unpaved roads; sources of secondary particles include reactions between gaseous chemicals emitted by power plants and automobiles.	150 Mg/m^3 (24-hour period for particles <10 µm); 35 µg/m^3 (24-hour period for particles <2.5 Mm).	Contributes to formation of haze as well as acid rain, which changes the ph balance of waterways and damages foliage, buildings, and monuments.	Irritation of breathing passages, aggravation of asthma, irregular heartbeat.
Lead (pb)	Metal processing, waste incineration, fossil-fuel combustion.	0.15 Mg/m^3 (rolling three-month average); 1.5 Mg/m^3 (quarterly average).	Loss of biodiversity, decreased reproduction, neurological problems in vertebrates.	Adverse effects upon mu - tiple bodily systems; may contribute to learning disabilities when young children are exposed; cardiovascular effects in adults.

The gaseous criteria air pollutants of primary concern in urban settings include sulfur dioxide, nitrogen dioxide, and carbon monoxide; these are emitted directly into the air from fossil fuels such as fuel oil, gasoline, and natural gas that are burned in power plants, automobiles, and other combustion sources. Ozone (a key component of smog) is also a gaseous pollutant; it forms in the atmosphere via complex chemical reactions occurring between nitrogen dioxide and various volatile organic compounds (e.g., gasoline vapours).

Airborne suspensions of extremely small solid or liquid particles called "particulates" (e.g., soot, dust, smokes, fumes, mists), especially those less than 10 micrometres (µm; millionths of a metre) in size, are significant air pollutants because of their very harmful effects on human health. They are emitted by various industrial processes, coal- or oil-burning power plants, residential heating systems, and automobiles. Lead fumes (airborne particulates less than 0.5 µm in size) are particularly toxic and are an important pollutant of many diesel fuels.

Except for lead, criteria pollutants are emitted in industrialized countries at very high rates, typically measured in millions of tons per year. All except ozone are discharged directly into the atmosphere from a wide variety of sources. They are regulated primarily by establishing ambient air quality standards, which are maximum acceptable concentrations of each criteria pollutant in the atmosphere, regardless of its origin.

Fine Particulates

Very small fragments of solid materials or liquid droplets suspended in air are called particulates.

Except for airborne lead, which is treated as a separate category, they are characterized on the basis of size and phase (i.e., solid or liquid) rather than by chemical composition. For example, solid particulates between roughly 1 and 100 μm in diameter are called dust particles, whereas airborne solids less than 1 μm in diameter are called fumes.

The particulates of most concern with regard to their effects on human health are solids less than 10 μm in diameter, because they can be inhaled deep into the lungs and become trapped in the lower respiratory system. Certain particulates, such as asbestos fibres, are known carcinogens (cancer-causing agents), and many carbonaceous particulates—e.g., soot—are suspected of being carcinogenic. Major sources of particulate emissions include fossil-fuel power plants, manufacturing processes, fossil-fuel residential heating systems, and gasoline-powered vehicles.

Carbon Monoxide

Carbon monoxide is an odourless, invisible gas formed as a result of incomplete combustion. It is the most abundant of the criteria pollutants. Gasoline-powered highway vehicles are the primary source, although residential heating systems and certain industrial processes also emit significant amounts of this gas. Power plants emit relatively little carbon monoxide because they are carefully designed and operated to maximize combustion efficiency Exposure to carbon monoxide can be acutely harmful since it readily displaces oxygen in the bloodstream, leading to asphyxiation at high enough concentrations and exposure times.

Automobile Emissions: Air pollution from vehicle tailpipes. A number of criteria pollutants, including carbon monoxide and nitrogen dioxide, are emitted by gasoline and diesel-powered vehicles.

Sulfur Dioxide

A colourless gas with a sharp, choking odour, sulfur dioxide is formed during the combustion of coal or oil that contains sulfur as an impurity. Most sulfur dioxide emissions come from power-generating plants; very little comes from mobile sources. This pungent gas can cause eye and throat irritation and harm lung tissue when inhaled.

Sulfur dioxide also reacts with oxygen and water vapour in the air, forming a mist of sulfuric acid that reaches the ground as a component of acid rain. Acid rain is believed to have harmed or destroyed fish and plant life in many thousands of lakes and streams in parts of Europe, the north-eastern United States, south-eastern Canada, and parts of China. It also causes corrosion of metals and deterioration of the exposed surfaces of buildings and public monuments.

Nitrogen Dioxide

Of the several forms of nitrogen oxides, nitrogen dioxide—a pungent, irritating gas—is of most concern. It is known to cause pulmonary edema, an accumulation of excessive fluid in the lungs. Nitrogen dioxide also reacts in the atmosphere to form nitric acid, contributing to the problem of acid rain. In addition, nitrogen dioxide plays a role in the formation of photochemical smog, a reddish brown haze that often is seen in many urban areas and that is created by sunlight-promoted reactions in the lower atmosphere.

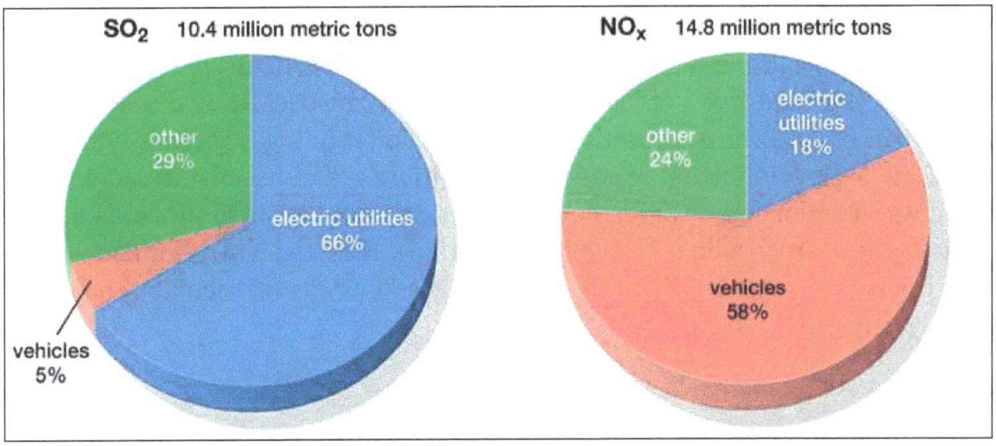

SO_2 and NO_x emissions in the U.S. SO_2 and NO_x emissions in the U.S.

Nitrogen oxides are formed when combustion temperatures are high enough to cause molecular nitrogen in the air to react with oxygen. Stationary sources such as coal-burning power plants are major contributors of this pollutant, although gasoline engines and other mobile sources are also significant

Ozone

A key component of photochemical smog, ozone is formed by a complex reaction between nitrogen dioxide and hydrocarbons in the presence of sunlight. It is considered to be a criteria pollutant in the troposphere—the lowermost layer of the atmosphere—but not in the upper atmosphere, where it occurs naturally and serves to block harmful ultraviolet rays from the Sun. Because nitrogen dioxide and hydrocarbons are emitted in significant quantities by motor vehicles, photochemical smog is common in cities such as Los Angeles, where sunshine is ample and highway traffic is heavy. Certain geographic features, such as mountains that impede air movement, and weather conditions, such as temperature inversions in the troposphere, contribute to the trapping of air pollutants and the formation of photochemical smog.

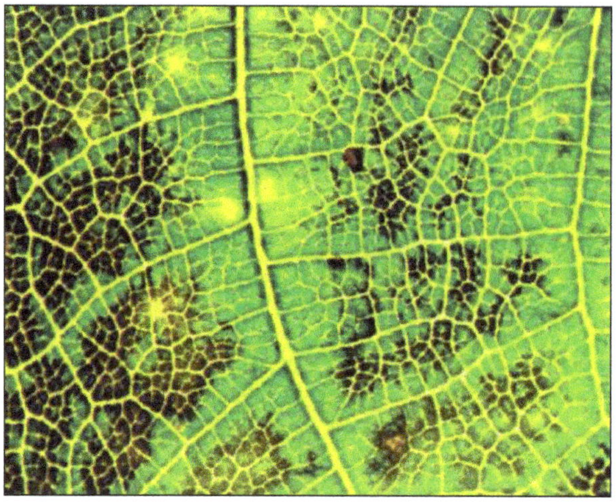

Ozone Damage On Leaf: Ozone damage on the leaf of an English walnut (Juglans regia).

Lead

Inhaled lead particulates in the form of fumes and dusts are particularly harmful to children, in whom even slightly elevated levels of lead in the blood can cause learning disabilities, seizures, or even death. Sources of airborne lead particulates include oil refining, smelting, and other industrial activities. In the past, combustion of gasoline containing a lead-based antiknock additive called tetraethyl lead was a major source of lead particulates. In many countries there is now a complete ban on the use of lead in gasoline. In the United States, lead concentrations in outdoor air decreased more than 90 percent after the use of leaded gasoline was restricted in the mid-1970s and then completely banned in 1996.

Air Toxics

Hundreds of specific substances are considered hazardous when present in trace amounts in the air. These pollutants are called air toxics. Many of them cause genetic mutations or cancer; some cause other types of health problems, such as adverse effects on brain tissue or fetal development. Although the total emissions and the number of sources of air toxics are small compared with those for criteria pollutants, these pollutants can pose an immediate health risk to exposed individuals and can cause other environmental problems.

Most air toxics are organic chemicals, comprising molecules that contain carbon, hydrogen, and other atoms. Many are volatile organic compounds (VOCs), organic compounds that readily evaporate. VOCs include pure hydrocarbons, partially oxidized hydrocarbons, and organic compounds containing chlorine, sulfur, or nitrogen. They are widely used as fuels (e.g., propane and gasoline), as paint thinners and solvents, and in the production of plastics. In addition to contributing to air toxicity and urban smog, some VOC emissions act as greenhouse gases and, in so doing, contribute to global warming. Some other air toxics are metals or compounds of metals—for example, mercury, arsenic, and cadmium.

In many countries, standards have been set to control industrial emissions of several air toxics. The fi st hazardous air pollutants regulated in the United States (outside the workplace environment) were arsenic, asbestos, benzene, beryllium, coke oven emissions, mercury, radionuclides (radioactive isotopes), and vinyl chloride. In 1990 this short list was expanded to include 189 substances as part of the significant amendments to the Clean Air Act of 1970. By the end of the 1990s, specific emission control standards were required in the United States for "major sources"—those that release more than 10 tons per year of any of these materials or more than 25 tons per year of any combination of them. Important air toxics, their sources, and their environmental effects are summarized in the table.

Important hazardous air pollutants (air toxics)		
Air toxic	Common sources	Human health risks
Arsenic	Volcanoes, weathering of rocks and minerals containing arsenic, industrial processes such as metal smelting, burning wood that has been treated with arsenic compounds.	Acute effects include gastrointestinal diso - ders, headaches, hypotension, shock, and death; chronic effects include dermatitis, irritation of mucous membranes, gastrointestinal disorders, and lung cancer.
Asbestos	Natural deposits, building materials.	Chronic effects include asbestosis (lung disease), pulmonary hypertension, lung cancer, and mesothelioma.
Benzene	Burning coal and oil, ve-hicle exhaust, industrial solvents, tobacco smoke.	Acute effects include dizziness, drowsiness, unconsciousness, and irritation of skin and mucous membranes; chronic effects include blood disorders including aplastic anemia, adverse effects on fetal development, and leukemia.
Beryllium compounds	Burning coal and oil, tobacco smoke, natural occurrence in soil.	Acute effects include inflammation and swelling of the lungs; chronic effect include chronic beryllium disease (development of noncancerous lesions in the lungs).
Cadmium compounds	Burning fossil fuels, incineration of waste, smelting of zinc, lead, and copper.	Acute effects include lung irritation; chro - ic effects include various kidney disorders, sperm-count reduction, and lung cancer.
Chlorine	Papermaking, drinking water, swimming pools.	Acute and chronic effects include irritation of mucous membranes and respiratory tract.
Coke oven emissions	Coke ovens.	Chronic effects include conjunctivitis, dermatitis, lesions of respiratory and digestive systems, and various cancers.
Cyanide compounds	Automobile exhaust, chemical processing, waste incineration.	Acute effects include headache, nausea, eye and skin irritation, and death; chronic effects include central nervous system disorders and thyroid disorders.
Formaldehyde	Building materials and home furnishings, power plants, incineration of waste, automobile exhaust, tobacco smoke.	Acute effects include irritation of eyes and respiratory tract and inflammation of mouth, esophagus, and stomach if ingested; chronic effects include lesions in respiratory tract, menstrual disorders, and lung and nasopharyngeal cancer.

Lead compounds	Battery manufacturing, paints and ceramic glaz-es, metal products.	Acute effects include gastrointestinal disorders and death; chronic effects include anemia, neurological disorders, reduced sperm count, and fetal effects including low birth weight and mental impairment.
Mercury compounds	Paint manufactured prior to 1990–91, batteries, industrial processes, dental amalgam fillings, fish products.	Acute effects include central nervous system effects including irritability, slowed sensory and motor nerve function, blindness, and deafness (depending on compound); chronic effects include a variety of central nervous system effects including erethism, insomnia, deafness, and blurred vision and kidney damage.
Nickel compounds	Nickel processing, fossil-fuel burning, sewage sludge incineration.	Acute effects include lung and kidney damage (anecdotal reports) and gastrointestinal disorders; chronic effects include dermatitis and asthma.
Radionucleides (radon, radium, uranium)	Ambient air (radon, uranium), drinking water (uranium), uranium mines and processing facilities, phosphate fertilizers.	Chronic effects include lung, bone, and nasal cancer.
Selenium compounds	Drinking water and ambient air, some plastics and paints, pharmaceutical production.	Acute effects include irritation of the resp - ratory and gastrointestinal tracts; chronic effects include loss of hair and tooth decay and discoloration.
Vinyl chloride	Manufacture of polyvinyl chloride (pvc), which is used in pipes, furniture, and automobiles.	Acute effects imclude central nervous system issues including dizziness, drowsiness, headaches, loss of consciousness, and inhibition of blood clotting; chronic effects include, rarely, vinyl chloride disease, in which changes occur in the bones and skin of the fingers and sensitivity to cold is increased, and liver cancer.

Pesticide Plant.

Air toxics may be released in sudden and catastrophic accidents rather than steadily and gradually from many sources. For example, in the Bhopal disaster of 1984, an accidental release of methyl isocyanate at a Union Carbide pesticide factory in Bhopal, Madhya Pradesh state, India, immediately killed at least 3,000 people, eventually caused the deaths of an estimated 15,000 to 25,000 people over the following quarter-century, and injured hundreds of thousands more. The risk of accidental release of very hazardous substances into the air is generally higher for people living in industrialized urban areas. Hundreds of such incidents occur each year, though none has been as severe as the Bhopal event.

Other than in cases of occupational exposure or accidental release, health threats from air toxics are greatest for people who live near large industrial facilities or in congested and polluted urban areas. Most major sources of air toxics are so-called point sources—that is, they have a specific location. Point sources include chemical plants, steel mills, oil refineries, and municipal waste incinerators. Hazardous air pollutants may be released when equipment leaks or when material is transferred, or they may be emitted from smokestacks. Municipal waste incinerators, for example, can emit hazardous levels of dioxins, formaldehyde, and other organic substances, as well as metals such as arsenic, beryllium, lead, and mercury. Nevertheless, proper combustion along with appropriate air pollution control devices can reduce emissions of these substances to acceptable levels.

Air Pollution: Air pollution from factory smokestacks.

Hazardous air pollutants also come from "area" sources, which are many smaller sources that release pollutants into the outdoor air in a defined area. Such sources include commercial dry-cleaning facilities, gasoline stations, small metal-plating operations, and woodstoves. Emission of air toxics from area sources are also regulated under some circumstances.

Small area sources account for about 25 per cent of all emissions of air toxics. Major point sources account for another 20 per cent. The rest—more than half of hazardous air-pollutant emissions—come from motor vehicles. For example, benzene, a component of gasoline, is released as unburned fuel or as fuel vapours, and formaldehyde is one of the by-products of incomplete combustion. Newer cars, however, have emission control devices that significantly reduce the release of air toxics.

Greenhouse Gases

Global warming is recognized by almost all atmospheric scientists as a significant environmental problem caused by an increase in levels of certain trace gases in the atmosphere since the beginning of the Industrial Revolution in the mid-18th century. These gases, collectively called greenhouse gases, include carbon dioxide, organic chemicals called chlorofluorocarbons (CFCs), methane, nitrous oxide, ozone, and many others. Carbon dioxide, although not the most potent of the greenhouse gases, is the most important because of the huge volumes emitted into the air by combustion of fossil fuels (e.g., gasoline, oil, coal).

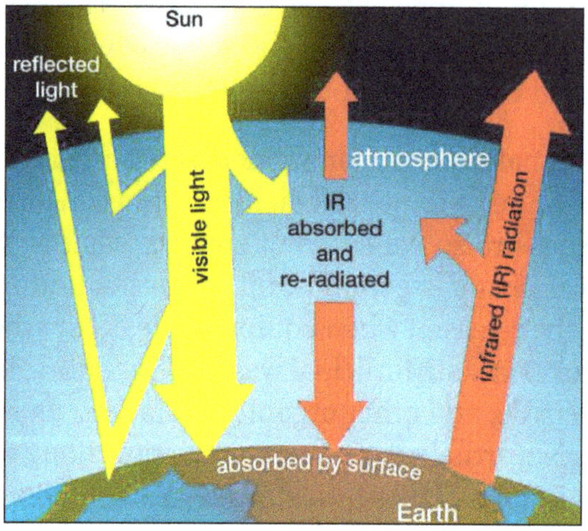

Greenhouse Effect On Earth

The greenhouse effect on Earth: Some incoming sunlight is reflected by Earth's atmosphere and surface, but most is absorbed by the surface, which is warmed. Infrared (IR) radiation is then emitted from the surface. Some IR radiation escapes to space, but some is absorbed by the atmosphere's greenhouse gases (especially water vapour, carbon dioxide, and methane) and reradiated in all directions, some to space and some back toward the surface, where it further warms the surface and the lower atmosphere.

Sources of Pollutants

There are innumerable sources of air pollutants, some known and some unknown. These sources can classified in many ways as mentioned below

- Natural Sources: The natural processes such as volcanoes, lightning, storms, etc, fall in this category.
- Anthropogenic Sources: Human activities constitute manmade or anthropogenic sources.
- Secondary Sources: The processes of formation of pollutants in atmosphere due to reactions of natural and manmade pollutants are secondary sources.

Pollutant sources have been categorized in following manner also:

- Stationary Sources: Such as homes, factories, power plants, etc.
- Mobile Sources: Such as vehicles, trains, aircrafts, ships, tractors, etc.

A further subdivision of pollutants as given below has also been proposed:

- Indoor Pollutant Sources: Such as kitchen.
- Outdoor Pollutant Sources: The sources outside the homes.
- Point Sources: Highly localized such as power plants, factories.
- Distributed Sources: Such as agricultural operations and cities.

- Local Sources: Such as power plants.
- Global Sources: Dust, soil, biological processes, volcanoes.
- Inorganic Sources: Soil, etc.
- Organic Sources: Agricultural processes, etc.

References

- Air-pollution-control, biology: byjus.com, Retrieved 10, Jan 2020
- Anthropogenic-air-pollution-sources, air-quality: intechopen.com, Retrieved 02, Feb 2020
- Common-iaq-pollutants: iaq.gov.hk, Retrieved 13, March 2020
- Indoor-nitrogen-oxides, advanced-air-pollution: intechopen.com, Retrieved 30, March 2020
- Air-pollutants-types-sources-effects-and-control-of-air-pollutants, environment-9964: yourarticlelibrary.com, Retrieved 17, April 2020
- Primary-pollutant, encyclopedias-almanacs-transcripts-and-maps, environment: encyclopedia.com, Retrieved 22, May 2020
- Air-pollution, science-286153: britannica.com, Retrieved 09, June 2020
- Major-sources-sulfur-dioxide-10011402: sciencing.com, Retrieved 22, July 2020

2

Atmospheric Aerosol: Size, Distribution and Dynamics

The microscopic particles of solid and liquid matter present in the air are commonly known as atmospheric aerosol. The process in which a small number of ions, atoms or molecules become arranged in a pattern like that of a crystalline solid and form a site upon which further particles are deposited is known as nucleation phenomenon. This chapter discusses in detail atmospheric aerosols and their related aspects.

Atmospheric aerosols are suspensions of tiny particles in air that originate from a wide variety of natural and anthropogenic sources. Depending on their sources, aerosols appear in different sizes and have diff rent physical and chemical properties. Aerosols play a significant role in the Earth's climate system and are a major contributor to air pollution. Despite considerable advances in recent decades, significant gaps still remain in our knowledge of aerosols and their impacts on our planet and human health. Aerosol is a system of liquid or solid particles uniformly distributed in a finely divided state through a gas, usually air

Aerosol particles, such as dust, play an important role in the precipitation process, providing the nuclei upon which condensation and freezing take place. They affect climate by reflecting or absorbing incoming solar radiation and enhancing the brightness, and thus reflectivity, of clouds. They also participate in chemical processes and influence the electrical properties of the atmosphere.

Aral Sea dust storm: A major dust storm occurring along the eastern coastline of the Aral Sea.

True aerosol particles range in diameter from a few millimicrometres to about 1 micrometre (equal to 10^{-4} cm). When smaller particles are in suspension, the system begins to acquire the properties of a true solution; for larger particles, the settling rate is usually so rapid that the system cannot properly be called a true aerosol. Nevertheless, the term is commonly employed, especially in the case of fog or cloud droplets and dust particles, which can have diameters of over 100 micrometres.

In general, aerosols composed of particles larger than about 50 micrometres are unstable unless the air turbulence is extreme, as in a severe thunderstorm. Particles with a diameter of less than 0.1 micrometer are sometimes referred to as Aitken nuclei.

Atmospheric aerosol particles originate from a wide variety of natural and anthropogenic sources. Aerosols are divided into two classes, namely primary aerosols and secondary aerosols, according to the mechanisms of their formation.

Primary aerosol: These particles are emitted directly into the atmosphere as liquid or solid form and come from combustion or fragmentation processes, for example- sea-salt, mineral aerosols (or dust), volcanic dust, smoke and soot, some organics, biological materials (plant fragments, microorganisms, pollen, etc.).

Secondary aerosol: The particles that are formed in the atmosphere by gas-to-particles conversion processes are secondary aerosols. The new particles are formed by nucleation and condensation of gaseous precursors for example- sulphates, nitrates, and some organics.

The aerosols are regularly form and introduced in the earth's atmosphere, undergo various physical and chemical interactions and transformations (such as coagulation, restructuring, gas uptake and chemical reaction etc.), and play a key role in a number of global processes such as:

- They serve as condensation nuclei and help in formation of clouds.
- They affect the abundance and distribution of atmospheric trace gases by heterogeneous chemical reactions and other multiphase processes.
- They can scatter, absorb and emit electromagnetic radiation.
- They serve as media upon which chemical reactions can occur.
- Airborne particles play an important role in the spreading of biological organisms, reproductive materials, and pathogens (pollen, bacteria, spores, viruses, etc.), and they can cause or enhance respiratory, cardiovascular, infectious, and allergic diseases.

A significant fraction of the atmospheric aerosols is anthropogenic in origin. Aerosols are responsible for climate change, visibility reduction, acid deposition and affect air quality and human health. Since the aerosols are highly important in the atmospheric phenomenon and affects the mankind in many ways, the study of their size and distribution become very useful to tackle the problems related with them.

Size Distribution

Primary and secondary aerosols are characterized by the size, shape, chemical content, amount and distribution in space and time, and how long they survive in the atmosphere. One of the most of important characteristics of poly-disperse aerosol is its particle size distribution, which represents the distribution of a specific aerosol property over the particle size range of interest. To construct a particle size distribution, the particle size is "weighted" by either number, mass, surface area, volume, or other aerosol property of interest. Normally shape of particles is assumed to be spherical for simplification of the mathematical problems related to the behaviour of aerosol particles.

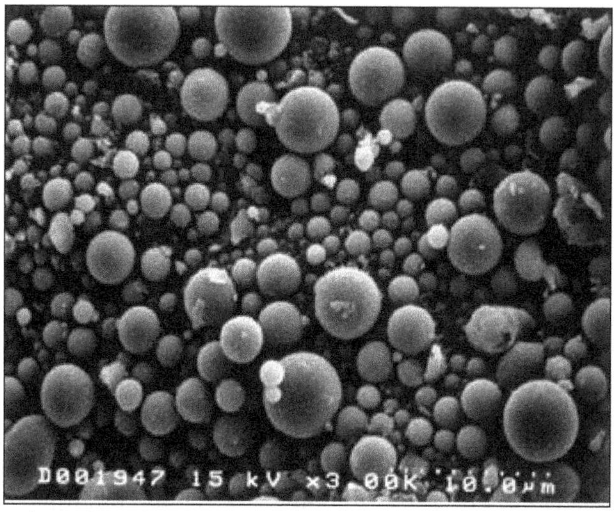

Photomicrograph made with a Scanning Electron Microscope (SEM): Fly ash particles.

Many aerosols comprise irregularly shaped particles. The non-sphericity of particles creates many problems. There exist also agglomerates of particles. If particle is non-spherical, its equivalent radius is introduced. There are several ways to define particle equivalent radius (for instance, aerodynamic equivalent radius, which is radius of a sphere that experience the same resistance to motion as the non-spherical particle). The diameters of atmospheric aerosol particles span over four orders of magnitude, from a few nanometers to around 100 μm. Particle number concentrations may be as high as 10^7 to 10^8 cm^{-3}. Thus, a complete description of the aerosol size distribution may be a challenging problem. Therefore, several mathematical approaches are used to characterize the aerosol size distribution.

The normal distribution has a fundamental defect related to its use in particle sizing analysis. This is because if the random variable (particle size) is normally distributed, it implies that the values of particle size are at equal distances from the central tendency.

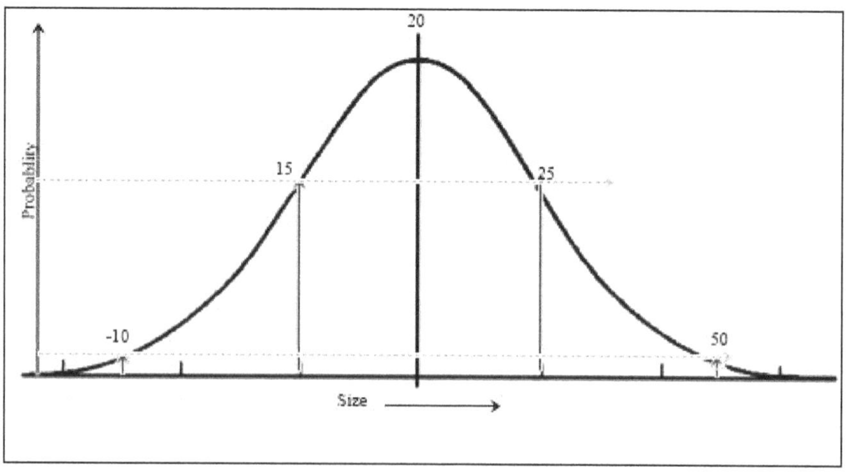

Normal distribution curve.

Now suppose the mean particle size or central tendency of the distribution is 20 μm. Then it would be equally probable to find either a 15 or 25 μm particle. One might also find a particle the size of 50 μm in the distribution and for a normal distribution, it would be equally likely to find a particle the size of minus 10 μm (not physically possible).

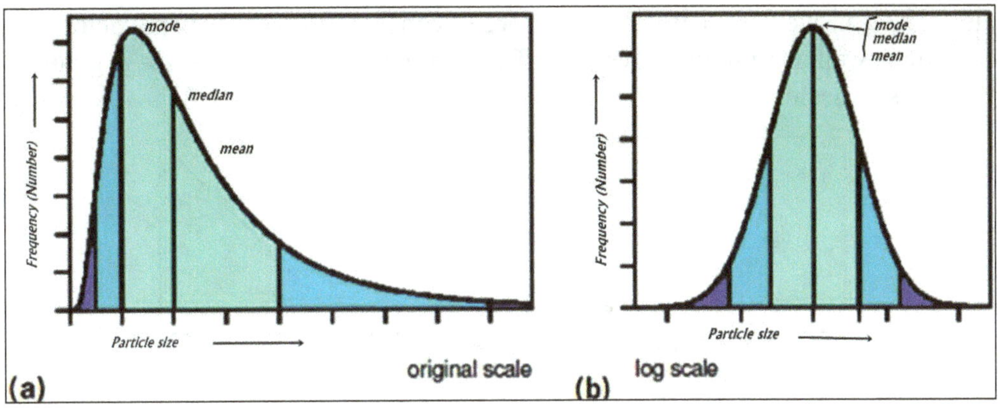

A log-normal distribution with original scale (a) and with logarithmic scale (b). Areas under the curve, from the median to both sides, correspond to one and two standard deviation ranges of the normal distribution.

In practical application, the size distribution of a typical dust is typically skewed to the right, i.e., skewed to the larger particle size. The central tendency of a skewed frequency distribution is more adequately represented by the median rather than by the mean. Statistically, the particle size distribution can be characterized by three properties: mode, median, and mean. The mode is the value that occurs most frequently. It is a value seldom used for describing particle size distribution. The average or arithmetic mean diameter is affected by all values actually observed and thus is influenced greatly by extreme values. The median particle size is the size that divides the frequency distribution into two equal areas.

When the skewed particle size distribution shown in Fig. (a) is replotted using the logarithm of the particle size, the skewed curve is transformed into a symmetrical bell shaped curve . This transformation is of great significance and importance in that a symmetrical bell-shaped distribution is amenable to all the statistical procedures developed for the normal or Gaussian distribution. In the lognormal particle size distribution, the mean, median, and mode coincide and have an identical value. This single value is called the geometric median particle size, d_g ,and the measure of dispersion is geometric standard deviation, σ_g . Thus the log-normal particle size distribution can be described completely by these two characteristic values. To determine whether the particles have a distribution close to lognormal distribution, the particle cumulative frequency data can be plotted on a logarithmic probability graph paper. If the distribution follows the lognormal relationship, then the plot will result in a straight line.

The geometric median particle size is the 50% value of the distribution as shown in Fig. The geometric standard deviation is equal to the ratio of 84.1% value divided by the 50% value or 50% value divided by the 15.9% value. Although 68.26% of the particles will lie in the particle size range between $(d_g + \sigma_g)$ and $(d_g - \sigma_g)$. If the particle size reduction is due to comminution such as crushing, milling, and grinding, the resulting particle size distribution very often tends to be log-normal distribution. The reason for the use of logarithmic or geometric size scale can be illustrated by considering an example of successive disintegration of a piece of blackboard chalk. For example, a 64 mm-long piece of chalk would break up into two pieces of 32 mm length each. Subsequent breakups yield pieces with lengths 16, 8, 4, 2, 1 mm, and so on, until one reaches molecular length scales. The ratio of adjacent sizes is always two, thus appearing at the same linear distance on a logarithmic or geometric size scale. Because with each breakage step, more and more particles are produced, the distribution is skewed, so that there are many smaller particles than larger ones.

This exercise of successive disintegration of a piece of chalk mimics the way particles are produced in many natural as well as industrial systems. Therefore, aerosol particle size is generally plotted on a logarithmic size scale.

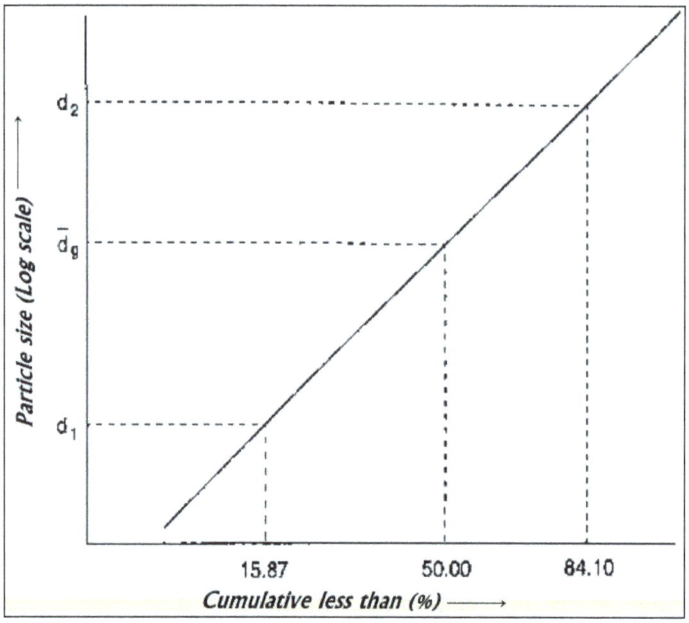

Log-normal distribution plotted on a logarithmic probability graph paper.

Log Normal Distribution

As explained above the atmospheric aerosols can be described well with a set of log-normal distribution functions (log-normal = normally distributed in logarithmic scale), which can be given by the following expression.

$$f = \frac{1}{\sigma_g \sqrt{2\pi}} \exp\left[-\frac{\left(z-\bar{z}\right)^2}{2\sigma_g^2}\right]$$

Where,

$$f = \frac{d\psi}{dz}$$

$$z = In\left(d_{pi}\right)$$

$$\bar{z} = \frac{\sum_i z d\psi}{\sum_i d\psi}$$

σ_g = geometric standard deviation

Ψ can be number, surface area, mass or volume.

Aerosol particle concentrations can be expressed by Number, Surface area, Volume, or Mass per unit volume. Based on particle distributions, different groups of atmospheric particles can be separated as:

- Nucleation (Aitken) mode: particle diameter < 0.1 μm
- Accumulation mode: particle diameter: 0.1 μm > d > 1 μm
- Coarse mode: particle diameter d > 1 μm

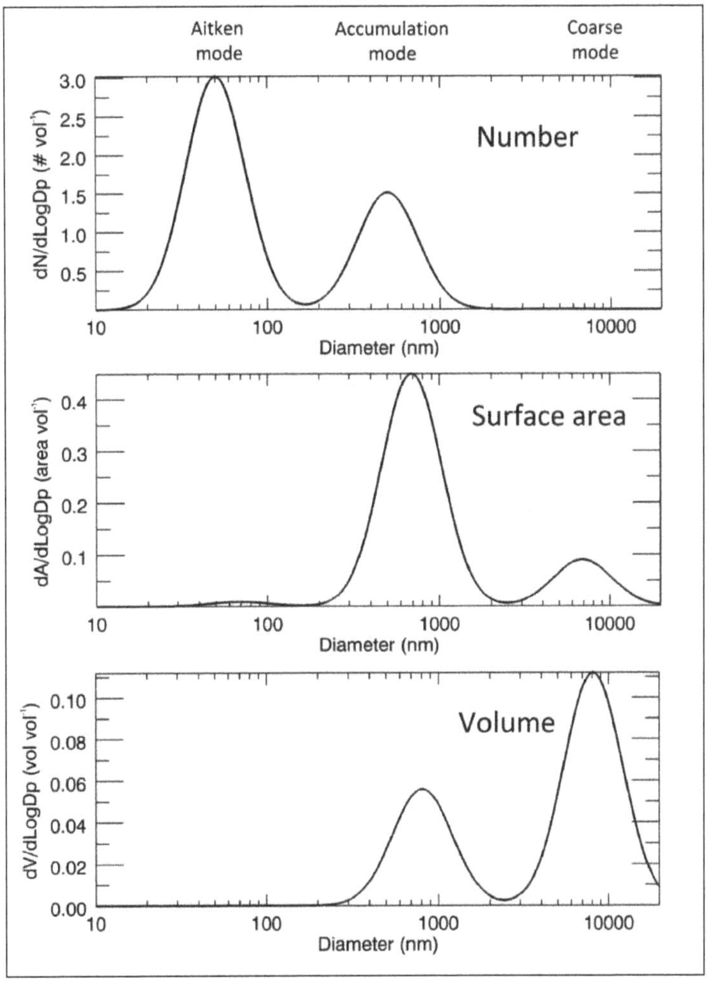

Lognormal distribution of Number, Surface area and volume versis particle size.

Aitken mode: Aerosol particles below 0.1 μm in diameter constitute the aitken mode. The smallest range of these particles (> 0.01 μm) sometimes called ultrafine mode. These particles are produced by homogeneous and heterogeneous nucleation processes. They can form during natural gas-to particle condensation or during condensation of hot vapor in combustion processes. Due to their rapid coagulation or random impaction onto surfaces, the lifetime of these small particles is very short (order of minutes to hours).

Accumulation mode: Larger aerosol particles in the size range 0.1 to 1 μm in diameter can accumulate in the atmosphere because their removal mechanisms are least efficient Their lifetime in the atmosphere is 7–10 days and during this period they can transported to a long distance from their sources. Particles belonging to this accumulation mode are formed mainly by coagulation of smaller particles or condensation of vapors onto existing particles, and during these mechanisms they growth into this size range. At the same time, they can also be emitted to the atmosphere from different sources, mainly from incomplete combustion. Accumulation particles removed from the atmosphere mainly by wet deposition.

Coarse mode: These contain particles with diameter larger than 1.0μm. These particles mostly emitted to the atmosphere during mechanical processes from both natural and anthropogenic sources (e.g. sea salt particles from ocean surface, soil and mineral dust, biological materials). Due to their relatively large mass, they have short atmospheric lifetimes because of their rapid sedimentation.

The number concentration is (in most cases) dominated by the ultrafine aerosols. Area concentration is dominated by the accumulation mode. The mass or volume concentration is dominated by the coarse and accumulation aerosols.

Aerosol Dynamics

Particles are emitted into the air, or may form in the air. Aerosol dynamics explains the evolution of complete aerosol populations. They spend some time in the air before being deposited from the air either naturally or artificially. The concentrations of particles will change over time as a result of many processes. External processes that move particles outside a volume of gas under study include diffu ion, gravitational settling, and electric charges and other external forces that cause particle migration. A second set of processes internal to a given volume of gas include particle formation (nucleation), evaporation, chemical reaction, and coagulation.

Here we shall look at some of the fundamental principles that control how particles behave under various conditions. Particles are deposited to surfaces by three main mechanisms: sedimentation, Brownian diffusion and impaction/interception

Drag Force

If a particle is moving through the air, or the air is moving past a particle, then a viscous drag force acts on the particle which acts opposite to its relative motion. Assuming that the flow of the particle is smooth (laminar) and not turbulent, then this drag force F_D is given by:

$$F_D = 3\pi\eta Vd$$

Where,

η = the gas viscosity. For air at tropospheric temperatures and pressures, η is around 1.8×10^{-5} Pa s

V= the relative speed between the gas and the particle

d= the particle diameter

We can determine whether the flow is smooth by calculating the Reynolds number (Re)

$$\mathrm{Re} = \rho_g \frac{Vd}{\eta}$$

Where, ρ_g is the gas density.

Flow starts to become turbulent for Re > 1, otherwise the flow is smooth. The drag equation assumes that the particle is large compared to the spaces between the molecules in the gas. This distance, known as the mean free path (λ), is about 0.07 μm for standard atmospheric temperature and pressure. As particle diameter decreases, there becomes an increased probability that the particle will move a distance without experiencing a collision, effectively slipping through the gaps between the molecules. Then we must consider the Cunningham slip correction factor C_C in calculating drag force. When d > 0.1 μm then Cunningham slip correction factor is given by:

$$C_C = 1 + \frac{2.52\lambda}{d}$$

Otherwise it is:

$$C_C = 1 + \left[2.34 + 1.05\exp\left(\frac{-0.39d}{\lambda}\right)\right]$$

Sedimentation

Every particle in the Earth's gravitational field experiences a force due to gravity towards the centre of mass of the Earth. If, for example, a stationary particle is released into the atmosphere, the gravitational force will accelerate it downwards. As its speed increases, the drag force will increase. The gravitational force acts downward, but the drag forces acts upward. The net force, which is the difference between the two, therefore decreases as the particle accelerates. Since the gravitational force remains constant, a speed will eventually be reached at which the two forces are equal and opposite and no more acceleration can take place. This is called the terminal or sedimentation speed (V_{term}). Particles that fall out of the air under the action of gravity do so at their terminal speed. For a spherical particle gravitational force (F_{grav}) is given by following expression:

$$F_{grav} = mg = \rho_{part}\frac{\pi d^3}{6}g$$

Where,

m = mass of particle

g = acceleration due to gravity

ρ_{part} = particle density

d = diameter of particle

When F_{drag} = F_{grav}, then we have

$$3C_C\pi\eta V_{term}D_{part} = \rho_{part}\frac{\pi d^3}{6}g$$

Hence,

$$V_{term} = \frac{\rho_{part} d^2 g C_C}{18\eta}$$

This equation is known as stokes' law which explains the sedimentation.

Brownian Diffusion

All the particles discussed here are very much larger than atmospheric molecules such as nitrogen and oxygen, and are continually bombarded all over their surfaces by such molecules. Brownian diffusion is the irregular movement of a particle immersed in a fluid caused by its collisions with the surrounding molecules of much smaller size. For very small particles, two effects become important. First, the rate of collisions with molecules becomes small enough that there is a reasonable probability of the collisions from one direction not being instantaneously balanced by collisions from the opposite direction. Second, the particle mass is small enough that the resultant momentum of these colliding molecules can exert a significant impulse in the unbalanced direction. So a small particle deflects slightly when gas molecules impact on them. The transfer of kinetic energy from the fast moving gas molecules to the small particle causes this deflection which is termed Brownian motion. Diffusi ity provides a measure of the extent to which molecular collisions cause these very small particles to move in a random manner across the direction of gas flow. The diffusion coefficien (D) in the equation below represents the diffusivity of a particle at given gas stream conditions

$$D = \frac{CkT}{3\pi\mu d_{pa}}$$

Where,

K = Boltzmann's constant

T = The absolute temperature

μ = The dynamic viscosity of air

dpa = particle aerodynamic diameter

C = Cunningham slip correction factor

Brownian diffusion becomes the dominant collection mechanism for particles less than 500 nm (0.5 mm) and is especially significant as the particles become smaller. Small particles obtain a high diffusion coefficien because it is inversely proportional to particle size. Thus, small particles in a fluid are subject to a random displacement known as the Brownian motion. This occurs in addition to the net motion in a given direction owing to the action of any of the external forces.

Coagulation

When there is relative movement of particles within a cluster by Brownian or turbulent diffusion, there is collision between particles which sometimes result in their combination to form a single new larger particle. This process is called coagulation. The new particle may be spherical if the original particles are liquid (e.g. sulphuric acid aerosol). But more likely it is untidy asymmetric clump of discrete parts, held together by electrostatic and molecular forces. Coagulation is fastest for smaller particles. The denser the particle cloud, the more will be collisions and coagulation. The basic rate equation is given by:

$$\frac{dN}{dt} = -K_0 N^2$$

Where,

$$K_0 = 4\pi d_{part} D = \frac{4kTC_C}{3\eta}$$

N = the number concentration of particles.

For air at ambient temperatures, K_0 is approximately 3 x 10^{-16} $C_c m^3 s^{-1}$. For a mono-disperse particle cloud with initial concentration N_0, the number concentration decreases with time according to:

$$N(t) = \frac{N_0}{1 + N_0 K_0 t}$$

Table: Concentration halving times and diameter doubling times for various initial concentrations. The value of K_0 is fixed at 5 × 1 $^{-16}$ $m^3 s^{-1}$.

Initial concentration No (counts/cm³)	Time to reach 0.5 No	Time for particle diameter to double
10^{14}	20 µs	140 µs
10^{12}	2 ms	14 ms
10^{10}	0.2 s	1.4 s
10^{8}	20 s	140 s
106	33 min	4 h
10^{4}	55 h	16 days
10^{2}	231 days	4 years

The particle diameter increases correspondingly, although at a slower rate because it takes eight original particles to produce one new particle of double the diameter. Equation for change in particle diameter is:

$$d(t) = d_0 (1 + N_0 K_0 t)^{\frac{1}{3}}$$

So by using above formulas it is possible to predict the change in number and size of particle with time.

Aerosols System and Nucleation Phenomenon

The aerosols are an important part of the atmosphere. The aerosol scatter the radiation from the sun, so impact the Earth's radiative budget and also act as condensation nuclei (CCN) for formation of cloud droplets. Aerosol particles are responsible for reducing visibility and impact the air quality. A major fraction of this important constituent of atmosphere is formed in the atmosphere through condensation of atmospheric vapours. There are several chemical compounds (nitric acid, organics, ammonia and water etc.) which can exist in the gaseous and aerosol phases in the atmosphere. When particles are formed the vapour is condensed into liquid or solid. The vapour molecules collide constantly with the liquid surface and become part of the liquid. At the same time there is a chance that molecules in the liquid escape from the liquid and become part of the vapour. At equilibrium these two processes balance.

So the concentration and particle properties keep changing with time. These changes can be the result of external forces, such as the loss of larger particles by gravitational settling, or they may be the result of physical and chemical processes that serve to change the size or composition of the particles. These physical and chemical processes involve mass transfer to from a particle. This transfer may be the result of molecular transfer between the particle and the surrounding gas, for example, condensation, evaporation, nucleation, adsorption, absorption, and chemical reaction, or it may result from inter-particle mass transfer, such as by coagulation. Processes that cause physical or chemical changes in the particulate phase influence the particle size distribution of nearly all aerosols

These processes contribute in an essential way to the earth's hydrological cycle. They are involved in the formation of photochemical smog and are the key to shaping the atmospheric aerosol size distribution. Condensation, thermal coagulation, and adsorption are related processes that rely on the diffusion of molecules or particles to a particle surface. Evaporation is the opposite of condensation and is governed by the same laws. Reactions may be non-growth processes that change the composition or density of an aerosol particle with little or no change in particle size.

Partial Pressure (p)

In a mixture of gases, each gas has its own partial pressure and it is equal to the pressure the vapour of a substance or gas would exert if it were the only component present at the same temperature. It describes the gas phase concentration of substance in a system.

Vapor Pressure (p_s)

It's also known as equilibrium vapor pressure or saturation vapor pressure. It is the pressure exerted by vapor in thermodynamic equilibrium with its condensed phases at a given temperature. It represents the minimum partial pressure of vapour of a liquid required to prevent the evaporation at the gas-liquid interface. So at saturation vapor pressure no net evaporation or condensation takes place.

Saturation Ratio (S_R)

It is the ratio of the partial pressure of vapor to the saturation vapor pressure.

$$S_R = \frac{p}{p_s}$$

Saturation ratio is very important for aerosol condensation and evaporation processes. When the saturation ratio is less than one ($S_R < 1$), then situation is sub saturated or unsaturated and evaporation takes place. But when the saturation ratio is greater than one ($S_R > 1$), then situation is supersaturated and condensation takes place. And when the saturation ratio is equal to one ($S_R = 1$), then situation is saturated and process is in equilibrium.

$S_R < 1$(sub saturated): evaporation

$S_R > 1$(supersaturated): condensation

$S_R = 1$(saturated): no net transport between the phases

The Kelvin Effect

The liquid aerosol particles have sharply curved surface so slightly greater partial pressure is required to maintain the mass equilibrium around aerosol as compared to the flat liquid surface at a given temperature. This increase in the partial pressure of vapor required for mass equilibrium, increases with decreasing particle size. This effect is called the Kelvin effect. The saturation ratio for such droplet having diameter d_p is given by the Kelvin equation:

$$S_R = \exp\frac{4\gamma M}{\rho_p RT d_p}$$

Where,

S_R = saturation ratio

γ = surface tension of liquid

M = molecular weight

ρ_p = density of the liquid

R = gas constant

T = temperature

d_p = diameter of droplet

Thus for smaller particle higher saturation ratio is required to prevent evaporation. For example, for 0.01 μm water droplets a saturation ratio of 1.24 is required (as compared to 1 for flat water surface) to prevent evaporation.

Nucleation

Nucleation is the source of new particles in the atmosphere. Nucleation is gas-to-particle conversion, in which low volatile gas phase species are converted to aerosol phases. Nucleation starts from the formation of sulfuric acid (H_2SO_4), because of its low vapor pressure. Formation of H_2SO_4 involves the following reactions:

$SO_2 + OH \rightarrow HSO_3$ (Rate limiting step)

$HSO_3 + O_2 \rightarrow HO_2 + SO_3$

$$SO_3 + H_2O \rightarrow H_2SO_4$$

SO_2 is emitted from fossil combustion and also forms from oxidation reactions of the biogenically emitted organic sulfur compounds. OH forms from the following atmospheric reactions:

$$O_3 \xrightarrow{h\gamma} O(^1D) + O^2$$

$$O(1D) + H_2O \rightarrow 2OH$$

Once formed, H_2SO_4 immediately becomes hydrated. The hydrated H_2SO_4 molecules are nuclei, and nucleation takes place by condensation of H_2SO_4 molecules and coagulation. That is, monomer becomes dimer, and dimer becomes trimer, and so on.

$$H_2SO_4 + H_2SO_4 \rightarrow 2\ H_2SO_4$$

$$2\ H_2SO_4 + H_2SO_4 \rightarrow 3\ H_2SO_4$$

$$3\ H_2SO_4 + H_2SO_4 \rightarrow 4\ H_2SO_4$$

$$4\ H_2SO_4 + H_2SO_4 \rightarrow 5\ H_2SO_4$$

The nucleation is of two types:

Homogenous Nucleation

The formation of the droplet from vapours is a complex process. Homogenous nucleation is the process of formation of droplets in the absence of any condensation nuclei. It takes place itself so it is also called self-nucleation. Homogenous nucleation needs large saturation ratios (in the range of 2 to 10), which is normally created in special laboratory or chemical process situations. Pure water vapour at 293 °K (200°C) and at a saturation ratio of 3.5 or greater spontaneously forms droplets by homogeneous nucleation. This corresponds to a Kelvin diameter of 0.0017 µm and suggests that molecular clusters of about 90 molecules are necessary for this process.

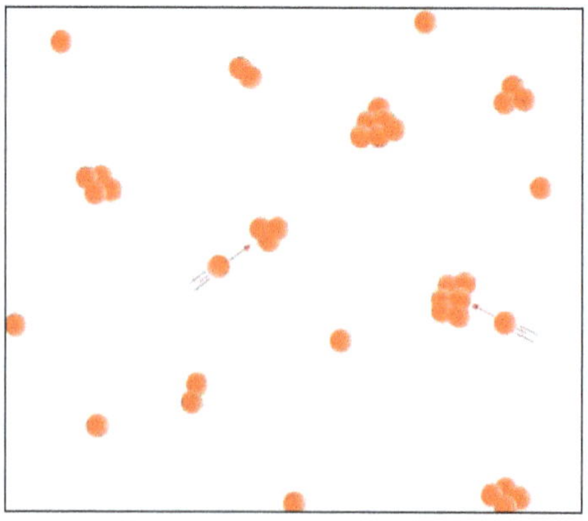

Homogenous nucleation.

Heterogeneous Nucleation

The other mechanism of nucleation is condensation or heterogeneous nucleation. This is common process and occurs on the existing sub-micrometre particles, called condensation nuclei. In the atmosphere there are thousands of these condensation nuclei per cubic centimetre of air which serve as sites for condensation. These nuclei could be insoluble or soluble.

Nucleation with Insoluble Nuclei

These nuclei serve as the passive sites on which condensation occurs. Under supersaturated conditions the vapour molecules get adsorbed on the wet table surface of solid (insoluble) nucleus. Now if the nucleus has a diameter greater than the Kelvin diameter for a given condition of super saturation, the vapor starts condensing on its surface. Once condensation starts, droplets start growing and rate of growth depends on:

- Saturation ratio
- Particle size
- Rate of arrival of vapor molecules at the droplet surface

In the beginning the droplet size is less than the mean free path of the surrounding gas $(d_p < \lambda)$, so the kinetic theory of gases governs the rate of arrival of vapor molecules at the droplet surface. Growth rate (cm/s) is given by equation:

$$\frac{d(d_p)}{dt} = \frac{2\alpha_c(p_\infty - p_d)}{\rho_p\sqrt{2\pi RT/M}}$$

Where,

α_c = condensation coeffic nt (the fraction of arriving molecules that stick)

p_∞ = partial pressure of vapor in the neighborhood of the droplet, but away from the droplet surface (dyne/cm²)

pd = partial pressure of vapor at the droplet surface (dyne/cm2)

M = molecular weight (g/mol)

ρ_p = density of the liquid (g/cm³)

R = gas constant (8.31 X 10^7 dyne.cm/K.mol)

T = temperature (°K)

When particle has grown in size more than the mean free path $(d_p > \lambda)$, then the rate of molecular diffusion governs the rate of arrival of vapor molecules to the droplet surface and growth rate is given by equation:

$$\frac{d(d_p)}{dt} = \frac{4D_vM}{\rho_p d_p R}\left(\frac{p_\infty}{T_\infty} - \frac{p_d}{T_d}\right)\varphi$$

Where

D_v = diffusion coefficient of the vapor molecu

∞ = subscript refers to conditions away from the particle, but in neighbour

d = subscript refers to conditions right at the droplet surface

φ = fuchs correction factor, it corrects for complications in the calculation of mass transfer by diffusion within one mean free path of the particle surface and is calculated as

$$\varphi = \frac{2\lambda + d_p}{d_p + 5.35\left(\frac{\lambda^2}{d_p}\right) + 3.42\lambda}$$

This factor can be omitted if the growing or evaporating droplet is > 2 μm and doing so offers a little error.

During rapid condensation (S_R > 1.05) the temperature of the droplet T_d will be greater than the surrounding air due to the release of heat of vaporization. The droplet temperature due to heating during condensation or cooling during evaporation can be given by equation:

$$T_d = T_\infty + \frac{(6.65 + 0.345T_\infty + 0.0031T_\infty^2)(S_R - 1)}{1 + (0.082 + 0.00782T_\infty)S_R}$$

Where, T_d and T_∞ are in °C. Partial pressure of vapor at the droplet surface (p_d) at temperature T_d is given by equation:

$$P_s = \exp\left(16.7 - \frac{4060}{T_d - 37}\right) \text{kPa}$$

Where T is in Kelvin.

Nucleation with Soluble Nuclei

This process is more complex and important. Our atmosphere contains a large number of droplets containing different soluble material. When water evaporates from these, solid residue left behind. For example the sodium chloride nuclei formed from sea spray. These soluble nuclei have a strong affinit for water which helps in the initial formation and growth of droplets even at lower saturation ratios than that needed for insoluble nuclei. So generally in presence of dissolved salt growth rate increases while evaporation rate decreases in comparison with pure liquid droplets. When a droplet containing dissolved salt grows or evaporates, two opposite effects work on the droplet. For instance, as a droplet evaporates its salt concentration increases, because only water evaporates. So this increases the water holding affinit of the dissolved salt in the droplet. The second effect is the Kelvin effect, i.e. with decrease in droplet size, an increase in the equilibrium vapor pressure (so higher saturation ratio) will be required for the droplet. The relationship between saturation ratio and particles size for droplets containing dissolved salts can be given by Kohler curves, shown below in figure

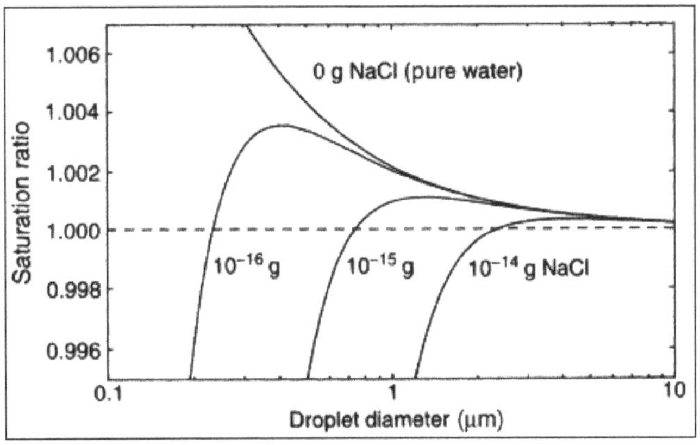

Saturation ratio versus droplet size for pure water and droplets containing the indicated mass of sodium chloride at 293 K (20 °C).

Figure shows three concentrations of dissolved salt (NaCl) and for droplet of higher concentration of dissolved salt lower saturation ratio is enough for its growth. The region below and the left of the peak for a given curve, represents equilibrium region, where droplet will remain at a given size until the saturation conditions change. Thus, there are a large number of particles in the atmosphere that will experience an increase in their size with an increase in relative humidity and a decrease with a decrease in relative humidity. The line for pure water does not have this type of equilibrium region. It represents only a demarcation between the growth (above) and evaporation (below) regions. As droplets continue to grow the concentration of dissolved salts decreases, eventually reaching the point where the droplets behave the same as pure water and their curves merge with that for pure water.

Nucleation Theory

Classical Nucleation Theory

Nucleation is a non-equilibrium process in which the system undergoes a phase transition from a metastable state to a stable state. Homogeneous gas phase nucleation is an important way of forming atmospheric nanoparticles from various gas phase species in the atmosphere. Due to its non-equilibrium nature, nucleation is not so well understood as other equilibrium processes, such as partitioning. The theoretical research on nucleation can be traced back to the pioneering work by Volmer and Weber, Becker and Döring, Frenkel, and Zeldovich in the early 1920-1940s. Based on their theory, the rate of nucleation is described by an Arrhenius type of equation:

$$J = J_0 e^{-} \beta \Delta G^*$$

Where,

J is the nucleation rate

J_0 is the pre-factor (the expression of J_0 as function of molecular properties can be found latter in the chapter)

ΔG^* is the Gibbs free energy change of critical nuclei formation

β is the inverse temperature

The value of ΔG^* can be derived from the following procedure. Generally, the formation of a cluster (A_n) containing n monomers from isolated n monomers (A_1) can be expressed as the following reaction scheme:

$$n A_1 = A_n$$

The Gibbs free energy change of $n A_1 = A_n$ is composed of two parts, bulk and surface contributions, which can be expressed as the following equation:

$$\Delta G(n) = -n(\mu_g - \mu_l) + \sigma_n S_n$$

where,

$\Delta G(n)$ is Gibbs free energy change of R1

μ_g and μ_l are the chemical potential of A_1 in gas phase and liquid phase, respectively

σ_n is the surface tension of cluster A_n

S_n is the surface area of cluster A_n

Assuming a spherical shape for the cluster A_n and within the capillary approximation, the following equations hold:

$$\sigma_n \equiv \sigma$$

$$n v_0 = \frac{4}{3}\pi R_n^3$$

$$S_n = 4\pi R_n^2$$

where,

σ is the surface tension of a flat surfac

R_n is the radius of cluster A_n

v_0 is the volume of monomer A_1

Substitute equation $\sigma \equiv \sigma$, $n v_0 = \frac{4}{3}\pi R_n^3$, $S_n = 4\pi R_n^2$ into equation $\Delta G(n) = -n(\mu_g - \mu_l) + \sigma_n S_n$, one obtains equation:

$$\Delta G(n) = -n\Delta\mu_{g-l} + \sigma\left(36\pi v_0^2\right)^{\frac{1}{3}} n^{\frac{2}{3}}$$

Thus, the formation free energy of cluster A_n is a function of cluster size n.

It is shown in Figure that there is a maximum of Gibbs free energy of formation $(\Delta G(n))$ along the size coordinate (n). Taking the derivative of equation $\Delta G(n) = -n\Delta\mu_{g-l} + \sigma\left(36\pi v_0^2\right)^{\frac{1}{3}} n^{\frac{2}{3}}$ and setting it to zero, the cluster size n* of maximum formation free energy can be easily obtained as shown in the following equations:

$$n^* = \frac{32\sigma^3\pi v_0^2}{3\Delta\mu_{g-l}^3}$$

$$\Delta G^* = \frac{16\pi v_0^2\sigma^3}{3\Delta\mu_{g-l}^2}$$

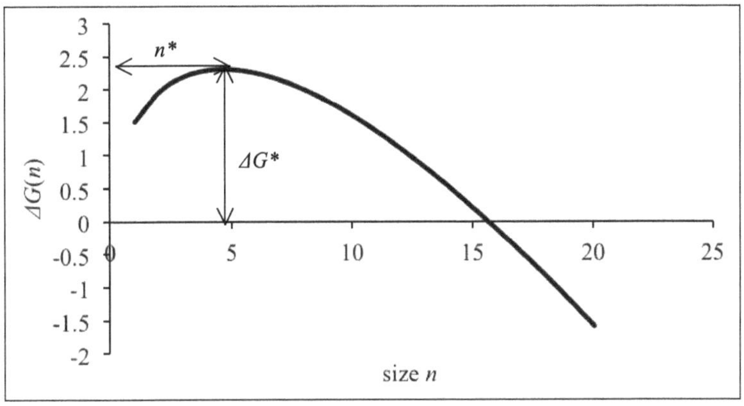

Plot of Gibbs free energy of formation against the cluster size.

Taking the derivative of equation $\Delta G^* = \frac{16\pi v_0^2\sigma^3}{3\Delta\mu_{g-l}^2}$ with respect to $\Delta\mu_{g-l}$, the following equation can be obtained:

$$\frac{\partial\Delta G^*}{\partial\Delta\mu_{g-l}} = -\frac{32\pi v_0^2\sigma^3}{3\Delta\mu_{g-l}^3} = -n^*$$

If J_0 is taken as a constant, combing equation $J = J_0 e^{-}\beta\Delta G^*$ and $\frac{\partial\Delta G^*}{\partial\Delta\mu_{g-l}} = -\frac{32\pi v_0^2\sigma^3}{3\Delta\mu_{g-l}^3} = -n^*$, the following equation can be obtained:

$$\frac{\partial \ln J}{\partial \ln S} = -\frac{\partial\left(\beta\Delta G^*\right)}{\partial\left(\beta\Delta\mu_{g-l}\right)} = n^*$$

Where, S denotes the super-saturation of the gas phase monomer A_1. The above equation $\frac{\partial \ln J}{\partial \ln S} = -\frac{\partial\left(\beta\Delta G^*\right)}{\partial\left(\beta\Delta\mu_{g-l}\right)} = n^*$ is commonly known as the first nucleation theorem

Kinetic Approach to Obtain the Pre-Factor J_0

It is impossible to obtain the pre-factor J_0 only from the above thermodynamic procedure. The following procedure has been described extensively in the literature. Considering nucleation from monomers A_1, the cluster A_n can be formed either by collision induced growth of small clusters $(A_{i<n})$ or by fission of large clusters $(A_{i>n})$. Since collisions involving three body or above are rare in the gas phase, it is reasonable to assume that all the collisions forming cluster A_n are two-body collisions. In addition, in the gas phase the concentration of monomer A1 is much larger than that of cluster $A_{i>1}$. As a result, the collision formation pathway of A_n can be further simplified as the collision between cluster A_{n-1} and A_1. If the major fission pathway of A_n is assumed to be the formation of A_{n-1} and A_1, the nucleation process can be denoted as the following reaction scheme reaction below:

$$A_1 \Leftrightarrow A_2 \Leftrightarrow A_3 \Leftrightarrow A_4 \Leftrightarrow \ldots \Leftrightarrow A_{n-1} \Leftrightarrow A_n \Leftrightarrow A_{n+1} \Leftrightarrow \ldots A_M$$

The flux J_n away from size n can be expressed as the following equation on the basis of reaction above:

$$J_n = \alpha_n C_n - \gamma_{n+1} C_{n+1}$$

Where,

C_i is the concentration of cluster A_i

α_i is the collision coeffici t α_i' times the monomer concentration C_1

γ_i is the evaporation coefficient of clust A_i

Assuming a constrained equilibrium is reached, the equation below holds:

$$0 = J_n = \alpha_n Z_n - \gamma_{n+1} Z_{n+1}$$

Where, Z_n is the constrained equilibrium concentration of cluster A_n.

Combine equation $J_n = \alpha_n C_n - \gamma_{n+1} C_{n+1}$ and $0 = J_n = \alpha_n Z_n - \gamma_{n+1} Z_{n+1}$,

$$J_n = \alpha_n C_n - \frac{\alpha_n Z_n}{Z_{n+1}} C_{n+1} = -\alpha_n Z_n \left(\frac{C_{n+1}}{Z_{n+1}} - \frac{C_n}{Z_n} \right) = -\alpha_n Z_n \frac{\partial}{\partial n} \left(\frac{C_n}{Z_n} \right)$$

Appling the continuity equation and equation above, the following equation holds:

$$\frac{\partial C_n}{\partial t} = J_{n-1} - J_n = -\frac{\partial J}{\partial n} = \frac{\partial}{\partial n} \left[\alpha_n Z_n \frac{\partial}{\partial n} \left(\frac{C_n}{Z_n} \right) \right]$$

The above equation is called Fokker Planck equation. Since nucleation is a non-equilibrium process, it is not surprising that it can be described by the Fokker Planck equation.

At steady state condition, the LHS of equation $\frac{\partial C_n}{\partial t} = J_{n-1} - J_n = -\frac{\partial J}{\partial n} = \frac{\partial}{\partial n} \left[\alpha_n Z_n \frac{\partial}{\partial n} \left(\frac{C_n}{Z_n} \right) \right]$ is zero and $J_n \equiv J$ for every n. Substitute $J_n \equiv J$ into equation $J_n = \alpha_n C_n - \frac{\alpha_n Z_n}{Z_{n+1}} C_{n+1} = -\alpha_n Z_n \left(\frac{C_{n+1}}{Z_{n+1}} - \frac{C_n}{Z_n} \right)$ $= -\alpha_n Z_n \frac{\partial}{\partial n}\left(\frac{C_n}{Z_n} \right)$, divide both sides by $\alpha_n Z_n$ and integrate, then:

$$J \int_0^M \frac{1}{\alpha_n Z_n} dn = \frac{C_0}{Z_0} - \frac{C_M}{Z_M} = 1$$

The last equality in equation $J \int_0^M \frac{1}{\alpha_n Z_n} dn = \frac{C_0}{Z_0} - \frac{C_M}{Z_M} = 1$ utilizes the boundary conditions $C_0 = Z_0$ and $C_M = 0$. As a result, the steady state nucleation rate can be calculate using the following equation:

$$J = 1 / \int_0^M \frac{1}{\alpha_n Z_n} dn$$

Based on equation $\Delta G(n) = -n\Delta\mu_{g-l} + \sigma\left(36\pi v_0^2\right)^{\frac{1}{3}} n^{\frac{2}{3}}$ and Boltzmann distribution, Z_n can be calculated using equation below:

$$Z_n = C_0 \exp\left[-\Delta G(n) / kT \right]$$

Since $\Delta G(n)$ has a maximum value at n*, the integration in equation $J = 1 / \int_0^M \frac{1}{\alpha_n Z_n} dn$ can be approximated using the steepest decent method:

$$J = 1 / \int_0^M \frac{1}{\alpha_n Z_n} dn = 1 / \int_0^M \frac{1}{\alpha_n . C_0 \exp\left(-\beta\Delta G^*\right) \exp\left[-\frac{1}{2}\beta \frac{d^2\Delta G(n^*)}{dn^{*2}} \left(n - n^*\right)^2 \right]} dn$$

$$= \alpha_n . C_0 \exp\left(-\beta\Delta G^*\right) / \int_0^M \exp\left[\frac{1}{2}\beta \frac{d^2\Delta G(n^*)}{dn^{*2}} \left(n - n^*\right)^2 \right] dn$$

Because $\frac{1}{2}\frac{d^2\Delta G(n^*)}{dn^{*2}} < 0$, the integration on the RHS of equation above can be evaluated using the gamma function. Therefore

$$J = \alpha_n . C_0 \exp\left(-\beta\Delta G^*\right) Z$$

$$Z = 1 / \int_0^M \exp\left[\frac{1}{2}\beta \frac{d^2\Delta G(n^*)}{dn^{*2}} \left(n - n^*\right)^2 \right] dn = \frac{\sqrt{-1/2\beta \frac{d^2\Delta G(n^*)}{dn^{*2}}}}{\Gamma(1/2)}$$

Where, Z is the commonly known Zeldovich factor in honor of J. B. Zeldovich who obtained the expression in 1942. Comparing equation above with equation $J = J_0 e^{-} \beta\Delta G^*$, the pre-exponential factor J_0 in $J = J_0 e^{-} \beta\Delta G^*$ can be obtained.

$$J_0 = \alpha_n . C_0 Z$$

Therefore, J can be expressed as the following equation,

$$J = \alpha_n . C_0 Z \exp\left(-\beta \Delta G^*\right) = \varpi_n . SC_{1,s} C_0 \exp\left(-\beta \Delta G^*\right) = \varpi_n . SC_{1,s} Z_{n^*}$$

Where, ω_n* is the collision rate; $C_{1,s}$ is the saturation concentration of monomer A1. Since Jo is not a constant with respect to lnS, the first nucleation theory in equation $\frac{\partial \text{In}}{\partial \text{In}} = -\frac{\partial(\ \Delta\ )}{\partial\left(\beta \Delta \mu_{g\ l}\right)} =$ must be modified accordingly

$$\frac{\partial \text{ In} J}{\partial \text{ InS}} = -\frac{\partial\left(\beta \Delta G^*\right)}{\partial\left(\beta \Delta \mu_{g-1}\right)} + \frac{\partial\ In J_0}{\partial\ In S} = n^* + 1$$

Recently, a more accurate expression for nucleation rate J was obtained by solving the equations $J_n = \alpha_n C_n - \frac{\alpha_n Z_n}{Z_{n+1}} C_{n+1} = -\alpha_n Z_n \left(\frac{C_{n+1}}{Z_{n+1}} - \frac{C_n}{Z_n}\right) = -\alpha_n Z_n \frac{\partial}{\partial n}\left(\frac{C_n}{Z_n}\right)$ and $\frac{\partial C_n}{\partial t} = J_{n-1} - J_n = -\frac{\partial J}{\partial n} = \frac{\partial}{\partial n}\left[\alpha_n Z_n \frac{\partial}{\partial n}\left(\frac{C_n}{Z_n}\right)\right]$ accurately. Fortunately, the new expression for J does not change the expression for first nucleation theorem

Nucleation Theorem for Multicomponent System

The theory for multicomponent nucleation can be obtained on the basis of generalizing the equations of mono-component nucleation.

The Fokker Planck equation $\frac{\partial C_n}{\partial t} = J_{n-1} - J_n = -\frac{\partial J}{\partial n} = \frac{\partial}{\partial n}\left[\alpha_n Z_n \frac{\partial}{\partial n}\left(\frac{C_n}{Z_n}\right)\right]$ and Boltzmann distribution equation $Z_n = C_0 \exp\left[-\Delta G(n)/kT\right]$ still hold in multicomponent nucleation. The counterparts of $\frac{\partial C_n}{\partial t} = J_{n-1} - J_n = -\frac{\partial J}{\partial n} = \frac{\partial}{\partial n}\left[\alpha_n Z_n \frac{\partial}{\partial n}\left(\frac{C_n}{Z_n}\right)\right]$ and $Z_n = C_0 \exp\left[-\Delta G(n)/kT\right]$ are shown in the following equations:

$$\frac{\partial C_n}{\partial t} = \nabla J = \nabla\left[\alpha_n Z_n \nabla\left(\frac{C_n}{Z_n}\right)\right]$$

$$Z_n = C_0 \exp\left[-\Delta G(n)/kT\right]$$

Where, $n = \left(n_1, n_2, \ldots, n_k\right)$ is the size space coordinate and $\nabla = \frac{\partial}{\partial n}$. In equation above, α_n is a n × n diagonal matrix attributable to condensation. Equation $Z_n = C_0 \exp\left[-\Delta G(n)/kT\right]$ can be expanded as a Taylor series at the saddle point (n^*) of the Gibbs free energy of formation $(\Delta G(n))$. If only the first three terms are used, Z_n can be expressed as:

$$Z_n = C_0 \exp\left[-\Delta G(n^*)/kT - \frac{1}{2}\beta(\Delta n)^T \underline{G}\Delta n\right]$$

Where, $\Delta n = n - n^*, \underline{G}$ is the Hessian matrix of Gibbs free energy of formation at n^*. Substitute equation above into $\frac{\partial C_n}{\partial t} = \nabla J = \nabla\left[\alpha_n Z_n \nabla\left(\frac{C_n}{Z_n}\right)\right]$, we obtain:

$$\frac{\partial C_n}{\partial t} = \nabla\left[\alpha_n.\exp\left(-\frac{1}{2}\beta(\Delta n)^T \underline{G}\Delta n\right)\nabla\left[\exp\left(\frac{1}{2}\beta(\Delta n)^T \underline{G}\Delta n\right)C_n\right]\right]$$

After transformation by changing of variables from n to v according to the following equations:

$$\Delta n = (\alpha_{n*})^{1/2} v, \frac{\partial}{\partial n} = (\alpha_{n*})^{-1/2}\frac{\partial}{\partial v}, \Gamma = \left((\alpha_{n*})^{1/2}\right)^T \underline{G}(\alpha_{n*})^{1/2}$$

equation $\frac{\partial C_n}{\partial t} = \nabla\left[\alpha_n.\exp\left(-\frac{1}{2}\beta(\Delta n)^T \underline{G}\Delta n\right)\nabla\left[\exp\left(\frac{1}{2}\beta(\Delta n)^T \underline{G}\Delta n\right)C_n\right]\right]$ can be written in the new coordinates v:

$$\frac{\partial C_n}{\partial t} = \frac{\partial}{\partial v}\left[\exp\left(-\frac{1}{2}\beta v^T \Gamma v\right)\frac{\partial}{\partial v}\left[\exp\left(\frac{1}{2}\beta v^T \Gamma v\right)C_n\right]\right]$$

Since Γ is a real symmetric matrix, it is diagonalizable and its eigenvalues are all real. It can be proved that Γ has and only has one negative eigenvalue and all the rest are positive due to the Hessian at saddle point. Assuming the negative eigenvalue is λ and the corresponding unit eigenvector pointing away from the origin is τ, equation above can be solved by separation of variables in the normal mode coordinate. At steady state the LHS of equation above equals 0. Therefore, in the direction of τ from (n*), C_n and J_n can be expressed as the following equations, respectively.

$$C_n = C_0 \exp\left(-\beta\Delta G(n^*)\right)\exp\left(\frac{1}{2}|\lambda|\beta\xi^2\right)\sqrt{\frac{|\lambda|\beta}{2\pi}}\int_{\xi}^{\infty}\exp\left(-\frac{1}{2}|\lambda|\beta x^2\right)dx$$

$$J_n = C_0\sqrt{\frac{|\lambda|\beta}{2\pi}}\exp\left(-\beta\Delta G(n^*)\right)(\alpha_{n*})^{1/2}\tau$$

In equation above, ξ is the coordinate in the direction of τ. Equation above shows that the direction of nucleation past from the saddle point is different from the steepest decent direction proposed by Reiss et al., since J_n is an eigenvector of $\alpha_{n*}\underline{G}$ instead of $\underline{G}$. If the number of component is 1, equation above will be reduced to equation $J = \alpha_n.C_0 Z\exp(-\beta\Delta G^*) = \varpi_n.SC_{1,s}C_0\exp(-\beta\Delta G^*) = \varpi_n.SC_{1,s}Z_n.$

There are also other simpler ways to derive the first nucleation theorem for multicomponent system. In the work by McGraw and Zhang, they showed that the first nucleation theorem took the following form:

$$(\frac{\partial \ln J}{\partial \ln S_i})_{sj,j\neq i,T} = n_i^* \delta_i, 0 \le \delta_i \le 1$$

Nucleation from Microscopic Point of View

The Gibbs free energy of formation determined from equation $\Delta G(n) = -n\Delta\mu_{g-l} + \sigma(36\pi v_0^2)^{\frac{1}{3}} n^{\frac{2}{3}}$ is based on macroscopic surface tension, which is not well defined as the cluster size approaches molecular scale. There are many studies focusing on obtaining the ΔG from microscopic properties without using the surface tension. One of the approaches is monomer density functional theory (DFT), which is an analogy for the commonly known DFT for electronic structure calculations. In the monomer DFT, the Gibbs free energy of formation (ΔG) is expressed as the functional of monomer spatial distribution density (ρ(r)). However, the intermolecular potential used in the monomer DFT is typically L-J or other simplified spherical potentials, as a result the monomer DFT is only applicable to the homogeneous nucleation of a small range of systems, such as Argon, Krypton, etc. More recently, the aggregation-volume-bias Monte Carlo (AVBMC) has been used to calculate the Gibbs free energy of formation of more realistic systems, such as water, acetic acid, and alkanes. In addition, the dynamical nucleation theory (DNT) has also been applied to calculate the nucleation rate. The main idea of DNT is to calculate the evaporation coefficien (γ_i) in equation $J_n = \alpha_n C_n - \gamma_{n+1} C_{n+1}$ using the variational transition state theory (vTST). The hard sphere model can be used to calculate the collision coefficien (α_i) in $J_n = \alpha_n C_n - \gamma_{n+1} C_{n+1}$. As a result, the equation $J_n = \alpha_n C_n - \gamma_{n+1} C_{n+1}$ for all i can be solved at steady state. At the critical size i*, evaporation coefficien (γ_i) and collision coefficien (α_i) are equal. The deficiency of DNT is that to calculate the steady state nucleation rate J, the evaporation rate of each cluster A_i needs to be obtained. As the cluster size increases, it requires too much calculation resources to obtain each γ_i, which makes the calculation impractical.

There are also some other attempts to obtain the Gibbs free energy of cluster formation using first principle calculations. This approach was widely applied to the nucleation system of atmospheric importance, such as the binary nucleation of sulfuric acid - water, ternary nucleation of sulfuric acid - water - ammonia/amines, and organic acids assisted nucleation. In these studies, the interaction energy between the nucleation precursors, such as sulfuric acid, water, amines, and organic acids, were calculated using the electronic density functional theory (DFT) and resolution identity Møller-Plesset perturbation (RI-MP2) theory. It is an advantage that these approaches do not need empirical intermolecular potentials and macroscopic surface tension. However, as cluster size increases the calculation burden increases dramatically and all the aforementioned studies failed to show the existence of a barrier in the nucleation coordinate. More efforts are needed to link the results obtained in the DFT and RI-MP2 calculations to the Gibbs free energy of formation of clusters in the future.

Coagulation

The aerosol growth also occurs through the collision of aerosol particles. This collision can be of two types and so is the coagulation i.e. thermal coagulation and kinematic coagulation.

Thermal Coagulation

The process where the collisions between the particles occur as a result of Brownian motion is called thermal coagulation. Thermal coagulation seems similar to the condensation, but it differs in many aspects. In thermal coagulation no super-saturation is required, there is no process like evaporation and on particle's surface other particle diffuses rather than molecule. This results in increase in average particle size and decrease in aerosol number concentration. But volume and mass concentration remain same if there are no addition or removal mechanisms.

Simple Monodisperse Coagulation: Aerosol particles exhibit Brownian motion and diffuse like gas molecules, but their diffusion occurs at a much slower pace; consequently, the diffusion coefficients for aerosol particles are about a million times smaller than those for gas molecules. The derivation developed by Smoluchowski gives the rate of change (decrease) in aerosol number concentration as:

$$\frac{dN}{dT} = -KN^2$$

Where,

N= particle number concentration (particles/m^3)

K= coagulation coefficient ^{3}/s)

For particles larger than the gas mean free path (For dp > λ), K is given by:

$$K = 4\pi d_p D = \frac{4kTC_c}{3\eta}$$

Where,

D = particle diffusion coefficien ^{2}/s

η = gas viscosity in Pa . s

K = Boltzmann's constant, 1.38 X 10^{-23} J/K

C_c = slip correction factor

In the usual situation, the extent of particle size increase is sufficientl limited that the coagulation coefficien can be considered a constant, and the rate of coagulation is proportional only to number concentration squared. Thus, coagulation is a rapid process at high number concentration and a slow one at low concentrations. The change in number concentration over a period of time t is given by:

$$N(t) = \frac{N_0}{1 + N_0 K_t}$$

Where,

N(t) = number concentration at time t

N_0 = initial number concentration

As number concentration decreases, particle size increases, but, for a closed system with no losses, total particle mass will remain constant. If number concentration decreases to one half of its original value, then the same mass (and volume) will be contained in half as many particles, so each particle will have twice its original mass (and volume). For liquid particles, particle size is proportional to the cube root of particle volume and, consequently, it is also proportional to the inverse cube root of number concentration:

$$d(t)=d_0\left(\frac{N_0}{N(t)}\right)^{\frac{1}{2}}$$

Thus an eight-fold reduction in particle number concentration results in a doubling of particle size. Combining above two equations we can give the change in particle size due to coagulation over a period of time t as:

$$d(t)=d_0\left(1+N_0K_t\right)^{\frac{1}{3}}$$

Table: Time Required for Selected Coagulation Processes.

Initial Number Concentration (m^{-3})	Time for Number Concentration to Halve (s)	Time for Particle Size to Double (s)
10^{18}	0.002	0.014
10^{16}	0.2	1.4
10^{14}	20	140
10^{12}	2000 (33 min)	14,000 (4 h)
10^{10}	200,000 (55 h)	1,400,000 (16 d)

Table gives the time required for various initial concentrations to reach one half their number concentrations and the time for particle size to double. It is apparent from Table that whether or not coagulation can be neglected depends on the concentration and time scale under consideration. Thus, over a period of a few minutes, coagulation is only important if particle number concentration exceeds $10^{12}/m^3$.

Polydisperse Coagulation: In the real world, we usually have a polydisperse aerosol, and the situation is more complicated. Because the coagulation process is governed by the rate of diffusion of particles to the surface of each particle, the process is enhanced when small particles with their high diffusion coefficients diffuse to a large particle with its large surface. A ten-fold difference in particle size produces a three-fold increase in coagulation rate. Now since in polydispersed coagulation the size of coagulating particles is different, so there will be a different value of K for them and has to be calculated separately. So average coagulation coefficient ($\bar{K}$) is be used in place of K to predict the change in number concentration over a period of time.

Kinematic Coagulation

In this type of coagulation the external forces creates the relative motion between the particles rather than the Brownian motion. In this the rate of coagulation increases with the particle number concentration. Because particles of different sizes settle at different rates, there is a relative motion between settling particles of different sizes. When particles travel at different velocities, the faster particles eventually overtake the slower ones, and particle contact occurs by interception if particles are big enough. This coagulation for particles moving in a flow velocity gradient is known as Gradient coagulation.

In turbulent flow, particles follow a complex path having strong velocity gradients. Relative motion between particles arises from these gradients and from inertial projection of the particles. The resulting coagulation is called turbulent coagulation. This mechanism is only important for particles larger than about 1 μm, and depends on the intensity the turbulence.

In acoustic coagulation, the relative motion between particles is created by intense sound waves. The smaller particles are more affected and oscillate with the sound waves. The relative motion that results leads to collisions and the process is called acoustic coagulation. Generally, sound pressure levels exceeding 120 dB are required to produce significant coagulation

Aerosols Sampling and Measurement Techniques

Sampling of environmental components is an important aspect of pollution control activities. Sampling of air and water is a systematic long-term measurement program of atmospheric pollutants and water contaminants, generally in specific locations. The primary reasons are

- To determine the concentration of pollutants.
- To determine the effects of pollutants on quality of life
- Assess long-term trends of pollutants.

Principles of Sampling and Analysis

The components of an air and water pollution monitoring system include the:

- Collection or sampling of pollutants both from the ambient air, water and from specific sources.
- The analysis or measurement of the pollutant concentrations.
- The reporting and use of the information collected.

Basic Considerations for Sampling

Sample must be representative in terms of time, location, and conditions to be studied. Sample must be large enough for accurate analysis. The sampling rate must be such as to provide maximum efficiency of collection. Duration of sampling must accurately reflect the fluctuations in pollution levels i.e. whether 1- hourly, 4-hourly, 6-hourly, 8-hourly, 24-hourly sampling. Continuous sampling is preferred. Pollutants must not be altered or modified during collection.

Sampling is the process of making individual measurements of pollutant concentrations. In passive sampling, a pollutant or pollutants is collected by devices that have no moving parts. In dynamic procedures (i.e., grab, intermittent, and continuous sampling), air is drawn through a device at a known rate using an air-moving system. A grab sample is a small volume of polluted air collected in a period of seconds to minutes. In intermittent sampling, air is collected for a distinct interval over which the concentration is integrated. In continuous sampling, realtime data is collected continuously 24 hours/day etc.

Two kinds of sampling instruments are available:

- Methods which collect samples and are analyzed in laboratory.
- Direct reading instruments capable of sampling a volume of air, immediate analysis and display reading.

The sample collection principles are:

- Absorption: A pollutant is collected by reaction with an absorbing reagent.
- Adsorption: A pollutant is collected on a solid surface.
- Impaction: Particles are collected on a surface by the use inertial principle.
- Filtration: Particles are collected on a ibrous medium as air passes through it.

During environmental sampling, it is also necessary to collect information on qualitative and quantitative data on the local sources of air pollution, topography, population distribution, land use pattern, climatology, etc, depending upon the objectives of the survey or measurement campaign. The location of the sampling site is dependent on the type of monitoring.

Sampling Location

Sampling locations are in general governed by factors like objectives, method of sampling and resources available.

Objective	Location
Health effect	breathing level in the population centres, hospitals, schools.
Vegetation	Foliage level.
Background concentration	Away from the sources of pollution.

Sampling Methods

Collection of gaseous samples:

- Adsorption
- Absorption
- Condensation
- Grab samples

Collection of particulate samples:

- Filtration
- Impingement
- Sedimentation

Aerosol

The environmental impact, chemical composition, concentration trends, and health effects of airborne particulate matter have been extensively studied and described in the literature. Current sampling methods involve the use of gravimetric filters or impactor devices and a wide variety of light and laser scattering devices. Many of the analytical methods for the determination of chemical composition of airborne particulate matter require sophisticated equipment and/or the use of strict sample preparation techniques. The tasks of sampling and analysis of airborne particulate matter are often complicated by the complexity of particle size, particle interactions, chemical partitioning between gaseous and particulate phase, and interactions with the sampling medium. The health effects of inhaled particulate matter are associated with both the size and shape of the particulate as well as its chemical toxicity.

There are two types of sampling – continuous and time averaged in–situ sampling. Continuous sampling is carried out by automatic sensors, optical or electrochemical, and spectroscopic methods which produce continuous records of concentration values. The specific time-averaged concentration data can then be obtained from continuous records. Time-averaged data can also be obtained by sampling for a short time – i.e. by sampling a known volume of air for the required averaging time. Samples are then analyzed by established physical, chemical, and biological methods for the concentration values which are the effective average over the period of sampling

Measurement of Aerosol Physical Properties

Integral Measurements

Instruments that provide integrals of specified variables over a given size range are often used for aerosol measurement. For example, condensation nucleus counters provide the total number of particles larger than a minimum size, and cloud condensation nuclei counters measure the subset of particles that can form cloud droplets when exposed to water vapor at a specified super saturation. Filter samplers are often used to measure total mass concentrations, integrated with respect to both size and time.

Number Concentration

Condensation nucleus counters (also referred to as condensation particle counters or Aitken Nuclei Counters: CNCs, CPCs, ANCs) measure the total aerosol number concentration larger than some minimum detectable size. Health effect studies have suggested that particulate health effects may be more sensitive to number concentration than to mass.

Cloud Condensation Nuclei Concentrations

Cloud condensation nuclei (CCN) counters measure the concentration of particles that are converted to cloud droplets by condensation of water (i.e., "activated") at a specified super saturation. CCN concentrations depend on both the aerosol size distribution and the aerosol composition.

Particulate Monitoring

Particulate monitoring is usually accomplished with manual measurements and subsequent laboratory analysis. A particulate matter measurement uses gravimetric principles. In this method, a filter-based high-volume sampler (a vacuum- type device that draws air through a filter or absorbing substrate) retains atmospheric pollutants for further laboratory weighing and chemical analysis.

Chemical analysis can be done by atomic absorption spectrometry (AAS), atomic fluorescence spectrometry (AFS), inductively couple plasma (ICP) spectroscopy, and X-ray fluorescence (XRF) spectroscopy, Ion Chromatograph(IC) and Carbon Analyser etc.

Measurement of Aerosol Chemical Composition

Measurements of particulate mass concentrations are important for regulatory and scientific reasons. There are two basic approaches to the measurement of aerosol particles. Aerosol measurements may be performed by collecting the particles onto a substrate followed by subsequent analysis. The collection surface may be a porous medium, e.g., a filter that passes the air or gas flow but retains all or a fraction of the particles suspended in the incident aerosol flow. Passage through such a porous medium generally has to be at a rather low flow rate and all of the aerosol flow must pass through extract this medium.

Measurement of particle composition typically involves the chemical analysis of deposited particles in a laboratory sometime after sample collection. Filters are the most commonly used collection substrates, but a variety of films and foils have been used with impactors to collect size-resolved samples. Sampling times vary with ambient loadings, sampling rates, substrate blanks, and analytical sensitivities but typically vary from several hours in urban areas to a day or more under clean background conditions.

Aerosols are determined by three parameters:

- Chemical Composition: What is the chemical makeup of the particulate matter?
- Mass Concentration: How much particulate matter is being inhaled by exposed persons?
- Size Characteristics: How big are the individual particles?

Off-line Measurements (Manual Method)

There are three types of particulate samplers: Filters, Impactors and Direct reading samplers.

Filter Sampling

	Pros	Cons
Glass FIber	• Low water uptake. • High efficiency for particles > 0.3 μ • Resistant to high T.	

Cellulose	• Low head loss. • Low metallic contamination level. • Easily ashed.	• Hydrophillic
Membrane	• Flat for microscopic analysis. • Hydrophobic. • Chemically resistant.	• Large head loss, • Brittle, • Not resistant high T.

The capture of aerosol particles by filters is the most common method of aerosol sampling. Air sample is aspirated through one or more openings in the solid casing of the filter holder. This method is a simple, versatile, and economical means for collecting samples of aerosol particles. A wide spectrum of filters, with markedly different physical and chemical properties, are available commercially for aerosol measurement applications in a variety of filter materials, pore sizes, and collection characteristics, as well as a variety of shapes such as discs and sheets sized to fit commonly available filter holders. Filters widely used for aerosol sampling may be classified according to their structure as fibrous filters, porous-membrane filters, Nuclepore filters, and granular-bed filters. Three types of Membrane filters are - Mixed cellulose ester (MCE), Teflon and Polyvinyl chloride. MCE filters are used for microscopy, particle counting and sizing and Fiber counting (asbestos). PVC filters have the advantage of being stable for gravimetric analysis. They do not absorb water vapor, on the other hand, teflon filters have low tare weight and low mass for gravimetric analysis. Filter cassettes of different diameter are available to place filters of 37mm, 47mm et

High Volume Sampler

Particle collection onto a filter is due to the simultaneous action of several collection mechanisms (e.g., diffusion, interception, inertial impaction, gravitational settling, and electrostatic attraction). The relative importance of these different mechanisms depends on the particle size, density, shape, and electrical charge, the filtration flow velocity and the mechanical and electrical properties of the lter.

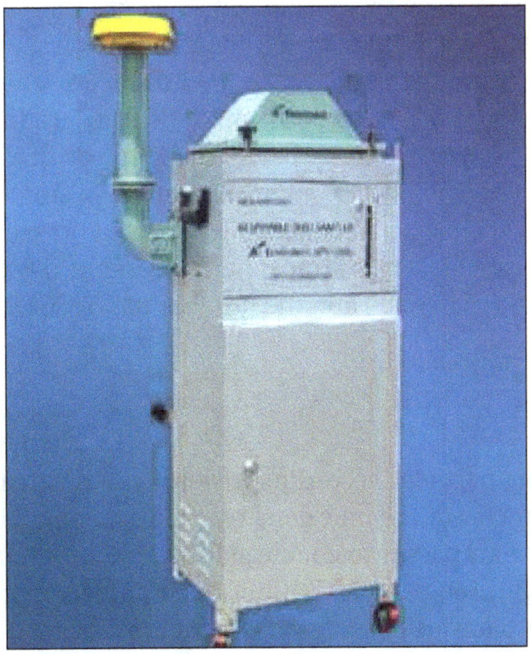

Advantages of HVS

- High flow rate at low pressure dro
- High particulate storage capacity
- No moisture regain
- High collection efficien
- Low coast
- No appreciable increase in air flo
- Resistance
- Filter is 99% efficient and can collect the particles as fine as 0.3
- Absorption principle is 99% efficient
- Collecting the gases

Inertial and Gravitational Collectors

Inertial and gravitational collectors include impactors, cyclones, aerosol centrifuges and impingers. In contrast to filters that generally collect particles of all sizes, these instruments distinguish between particle sizes. They are used for size-selective sampling or size-segregated aerosol collection of airborne particles depending on the aerodynamic diameter of the particle. Cyclones and single-stage impactors are used as a pre-cut for size-selective sampling and are commonly followed by a filter for aerosol collection

Each stage of a cascade impactor can be analyzed to determine aerosol mass distributions or to assess chemical composition as a function of particle size. Cascade cyclones are useful under elevated temperature conditions such as those found in stack gas sampling. Aerosol centrifuges and some horizontal elutriators size-fractionate over a continuous size spectrum and are used to determine aerodynamic shape factors for irregularly shaped particles. The choice of sampler depends upon the application and the analyses to be performed. Because traditional impactors do not provide much size resolution for sub micrometer particles, micro-orifice and low-pressure impactors have been developed to obtain smaller cut-point diameters as small as 0.05 μm. Low-pressure impactors resemble ordinary impactors but operate at reduced pressures of 0.05 to 0.4 atm, while micro-orifice impactors operate closer to atmospheric pressure 0.8 to 0.9 atm but employ very small orifices (40 to 200 μm in diameter) and the number of nozzles is large to obtain an adequate flow rate

Impactors

Impactors are used to classify particles according to aerodynamic diameter. The aerodynamic diameter is defined as the diameter of the unit density sphere having the same settling speed as the particle. In Cascade impactor air passes from one stage to the next to remove particles in discrete size ranges, and is exhausted through an after-filter. Each impactor stage consists of one or more jets followed by a collection plate and successive stages are designed to collect smaller particles.

Those particles, which penetrate all the impaction stages, are collected by a final absolute filter. The size collected by any individual stage depends upon the jet diameter and the air stream velocity in the jet as well as distance to the impaction surface; therefore, the collection of smaller particles in subsequent stages is achieved by using a combination of: smaller diameter jets, higher jet air velocities, or decreased distance to the impaction surface.

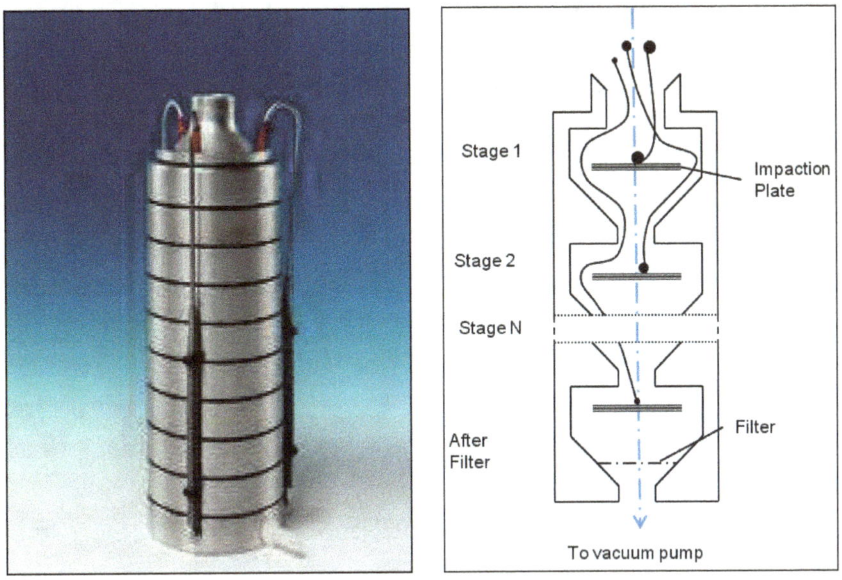

Cascade Impactor.

The major limitation of impactors is particle bounce from the collection surface and re-entrainment because of the continuous blowing of the jet on the particle deposit; however, this is minimized by using a sticky surface on the impaction plate, e.g. numerous types of greases and oils. Impactors may be multistage or single stage and multiple fractions of varying size ranges may be collected. They can be used to determine complete size distribution. Impactors are flow rate dependent and are used for bioaerosol assessment. Mostly area-type measurements are done because high flow rate is required

Cyclones

Cyclones are commonly used size-selective samplers. They are used to define respirable particle sampling criteria. Cyclone removal mimics respiratory system when operated. Flow rate is critical. Cyclones use centrifugal forces for particle collection and utilize a vertical flow inside a cylindrical or conical chamber. Particles with sufficien inertia are unable to follow the air streamlines and impact onto the cyclone walls. The particles are either retained on the cyclone walls or migrate to the bottom of the cyclone cone, while finer particles remain entrained in the air during the upwards spiral.

Cyclones have several advantages in air sampling including their large capacity for loading and insensitivity to orientation. Furthermore, they have relatively low cost of construction, ease of operation, and are easily maintained. Unlike impactors, they are not subject to errors due to particle bounce and re-entrainment and do not require special collection surface coatings. However, the flow pattern inside cyclones is complex and not easily modeled; thus, it is not easy to predict cyclone performance without reference to empirical correlations.

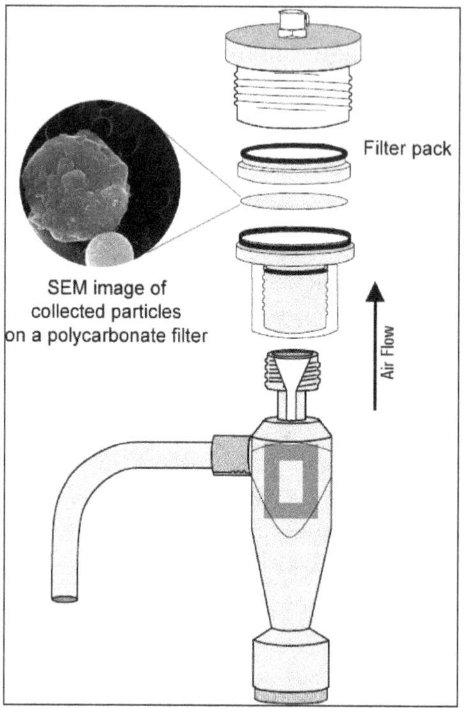

Cyclone Sampler.

Impingers

Impingers utilize inertial properties of particles to effect collection and operate much like impactors, except that the sampled air stream jet is immersed in water at the bottom of a flask. The sampled air stream is accelerated in the impinger orifice to velocities of 60 m/s or greater. The air stream exits underneath the liquid surface immediately above an impaction plate or at a specified distance above the bottom of the collection flask. Particles impinge on the plate or flask bottom, stop, and are subsequently retained by the liquid. Impingers are effective for the collection of particles in the 1μm to 20μm size range. The lower size collected depends on jet velocity and diameter. The upper size collected by impingers is limited because large particles cannot follow the air stream into the impinger.

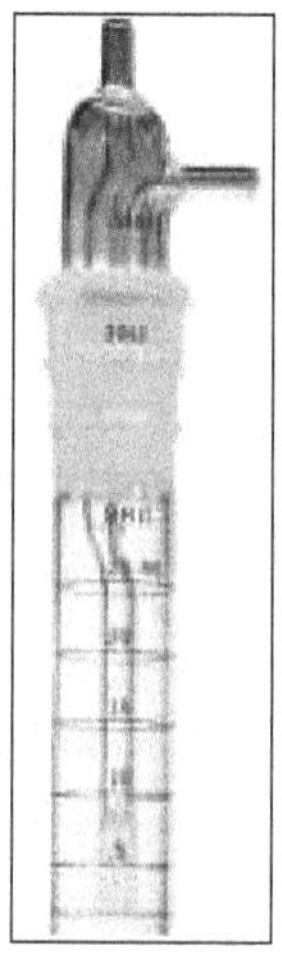

Impinger.

Elutriators

Elutriators use gravitational settling in a laminar flow to separate particles by aerodynamic diameter. They provide segregation for particles greater than 3µm. Two types of elutriators, vertical and horizontal, are available. The vertical elutriator consists of a vertical duct through which air flows slowl upward. Particles whose sedimentation velocity is greater than the duct velocity cannot follow the air flow and settle out. The ability of the device to distinguish particle sizes is reduced by the distribution of velocities in the duct; nonetheless, it is an effecti e method for removing large particles.

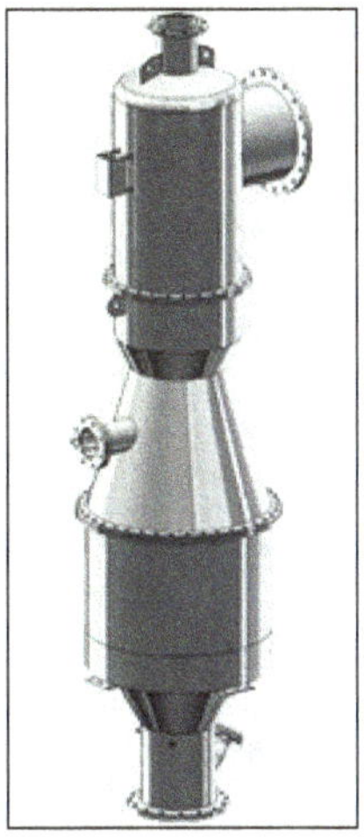

Elutriator.

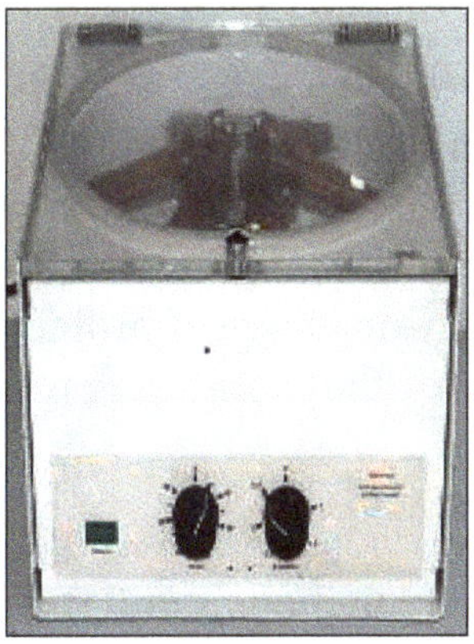

Aerosol Centrifuge.

Aerosol Centrifuges

Aerosol centrifuges refer to a class of aerosol samplers that spin at high velocity in order to subject particles to large centrifugal force which is used to deposit the particles on the outer edge of an aerosol chamber. Although several centrifuges have been developed, spiral centrifuge is the one that has seen the greatest application. Generally, particles are collected on a foil which lines the outer wall of the channel and is removed for particle analysis after collection where the deposition distance along the foil is directly related to the aerodynamic diameter of the particle. The resolution of the centrifuge depends on the ratio of the aerosol sampler flow rate to the total flow rate and the overall particle classification may be achieved within the range $0.01 < d < 5$ µm at sampling flow rates up to about 1 l/min

Real-time Measurement Methods

In real-time or direct-reading instruments, sampling and analysis are carried out within the instrument and the property of interest can be obtained almost immediately. Different aerosol properties, such as number concentration, mass concentration, opacity, size distribution, etc. can be measured. The inorganic constituents of $PM_{2.5}$ (Na^+, NH_4^+, K^+, Mg^{2+}, Ca^{2+}, Cl^-, NO_3^-, SO_4^{2-}) and associated precursor gases (HCl, HNO_3 and NH_3) are measured by the Ambient Ion Monitor - Ion Chromatograph. The AIM-IC system provides a Continuous, near-real-time, hourly online measurement of the watersoluble inorganic constituents of $PM_{2.5}$ and precursor gases.

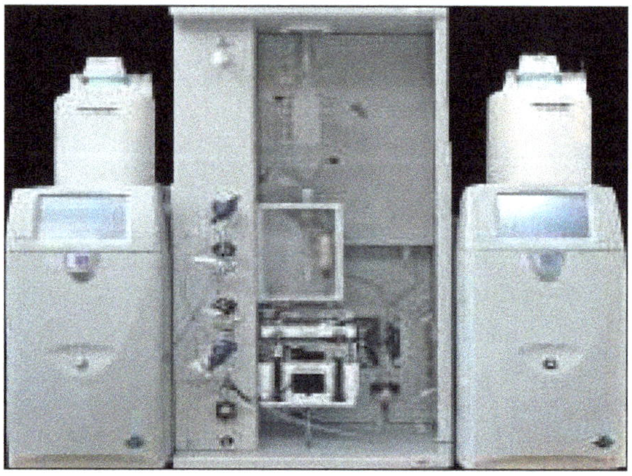

Ambient Ion Monitor - Ion Chromatograph.

Gas Sampling

The reduction of pollution and human health damages requires an integrated approach to air and water quality management. An important step in developing a management strategy is to be able to monitor and evaluate air and water quality. A good monitoring and modeling system is essential for policy making suited to the primary objective of protecting human health. Choice of samples and sampling methods are important for the monitoring results to be significant

Sampling of Gases and Vapors

The term gases and vapors are used synonymously and the methods employed for their monitoring are same. Gases are substances which are non-condensable at room temperature and under ordinary conditions exist in the gaseous state even when present at high concentrations. Vapors are derived from volatile liquids and may condense at high concentrations. A substance in aerosol or particulate form may coexist with its vapor. The methods applied for sampling gases and vapors are however similar.

Sampling Procedures

Gaseous analysis can be conducted by collecting the sample at the site and performing the analysis in the laboratory or by employing direct reading instruments which sample a known volume of air, perform the analysis immediately and display the results visually. In methods involving laboratory analysis of the samples, a known volume of air containing the contaminant of interest is collected over a short time period usually a few seconds to few minutes. Such a sample is called as Grab sample and the results of the analysis are representative of the contaminant concentration at that sampling site at that time. Grab sampling method is usually employed to monitor different phases of a cyclic process or to monitor and determine the peak concentrations of a contaminant emitted from an industrial process and whose concentrations vary with time. Grab sampling can be performed by the use of various gas sampling devices like evacuated flasks, bottles, metallic cylinders, syringes, plastic bags and gas-liquid displacement containers. Grab sampling does not require air flow measurements, unlike the conventional methods involving adsorption or absorption of the gas in liquid and solid mediums.

The volume of sample collected is measured directly from the volume of the container or a known volume is withdrawn with the help of syringes and directly injected into the injection port of the analytical instrument. It is however necessary to record the temperature and pressure during the sampling and correct them to standard conditions while reporting. The collection efficienc of grab sampling is generally 100%, however sample decay can occur from sorption or sample decay. The availability of modern instruments makes this method highly sensitive. Vinyl chloride, for example, is measurable in grab samples by gas chromatography at levels well below 1.0 ppm. Grab sampling is not preferred to collect reactive gases like H_2S, oxides of nitrogen and sulfur oxides, which can react with the dust particles and moisture in the sample or react with the sealants and material of the grab sampler and alter the chemical composition of the sample. In such cases it is advisable to analyze the sample immediately after collection.

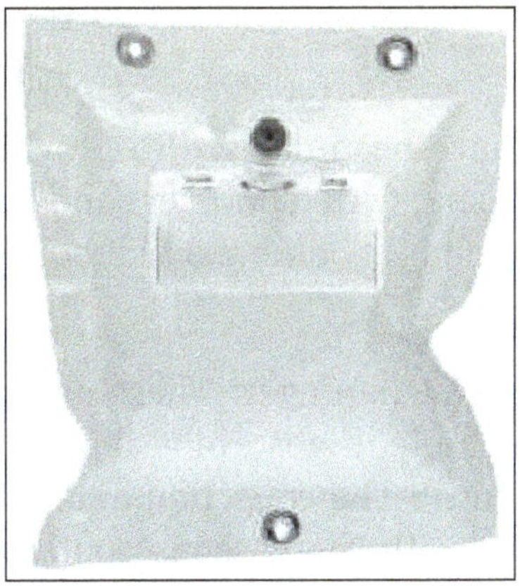

Teflon bag Grab sampling Devices

Stainless steel canister.

The second method involves collection of an Integrated sample (average or long term) in which a known volume of air is passed through an absorbing or adsorbing media to remove the contaminant/gas from the air stream. The sampling period may vary from duration of less than an hour to eight hours. Analysis of these samples yield integrated, average or long term exposure levels. Integrated sampling is employed when the concentrations of gaseous contaminant varies over a short duration, or when the concentrations of the contaminant are low and extended sampling durations are required to meet the sensitivity requirements of the analytical method or to assess the exposure levels of a contaminant in work places in order to establish compliance non-compliance with eight hour or timeweighted exposure health standards.

Absorption

In this method the gaseous pollutant is extracted from the air stream by passing it through a liquid collecting phase, whereby the gas molecules either preferentially dissolve in the liquid phase or may chemically react with it. The extent of absorption of a given gaseous contaminant in a particular solvent depends on its solubility and is limited by the equilibrium partial pressure of the vapor over its solution. The efficienc of absorption of a gas into the liquid phase also depends on the contact surface area hence the absorbing devices are designed to offer maximum contact between the gas bubbles and the absorbing media. This is achieved by transforming the air stream to small finely dispersed bubbles with a relatively long travel time through the absorbent. Solubility of a gas in a particular solvent also depends on temperature; lower temperatures favour the absorption of the more volatile constituents.

Four types of absorption devices have been devised to provide efficien absorption: simple gas wash bottles, spiral or helical absorbers, fritted bubblers and glass beaded columns. The selection of the absorber depends upon the solubility and reactivity of the gaseous contaminant to be collected.

1. Simple gas wash bottles and midget impinger: These are used for collection of non-reactive gases and vapors that are highly soluble in absorbing liquids. They have the advantage of high flow rate capacity upto 30 litres min^{-1}. The collection efficiency can also be improved by using absorbing reagents which can react with the gaseous contaminant and forming a more stable compound. Examples of these are 2,4-toluenediisocyanante (TDI) hydrolyzed by an absorbing reagent solution to a toluenediamine derivative and p,p-diphenylmethane diisocyanate (MDI) hydrolyzed to methylene dianiline. Midget impingers are commonly employed in sampling of gases and vapors in occupational environments. However they can be used to collect general area air samples from a stationary position; they can be hand–held to collect breathing zone samples or they can be attached to a workers clothing for a personal sample. The efficiency of collection can be enhanced by entraining two or more impingers in series. These impingers have the disadvantage of spillage often encountered during field sampling. However, spill proof impingers have been developed and are commercially available.

2. Spiral Absorbers: These absorbers are used to collect gaseous contaminants that are moderately soluble in the absorbing reagent or react with it slowly. The spiral/helical structure increases the collection efficiency as the air has to travel a spiral or helical path through the liquid which allows a longer residence time within the tube and results in longer contact between the sampled air and the absorbing medium.

3. Fritted bubblers: These are more efficient absorbing devices commonly used for sampling of gaseous contaminants that are only slightly soluble or reactive with the absorbing medium. These are usually employed for sampling gaseous contaminants in ambient work atmospheres. The air sample is drawn through a sintered or fritted glass bubbler which is submerged in an absorbing solution or reagent. As the sampled air is drawn through the bubbler small bubbles and a heavy froth develops, increasing the surface area and contact time between the gaseous contaminant and the absorbing solution. The bubble size depends upon the nature of the absorbing liquid and diameter of the orifices from which the bubbles emerge. Frits can be fine, coarse and extra coarse depending on the number of orifices per unit area. Coarse frits are used to sample a gaseous contaminant that is appreciably soluble or reactive with the absorbing medium. Frits of medium porosity are used to sample gaseous contaminants that are difficul to collect while fine porosity frits are used to collect highly volatile gaseous substances.

If concentrated solutions of a gaseous contaminant are required then sample collection is performed over columns packed with glass beads. The beads provide a large surface area for collection. This technique has been employed for collection of benzene and other aromatic hydrocarbons.

Adsorption

Adsorption is a common air sampling method used to collect trace quantities of insoluble and nonreactive gases and vapors on solid adsorbents. Adsorption is a surface phenomenon whereby gases are concentrated and bound by intermolecular forces to the adsorbent phase. Under equilibrium conditions at constant temperature the volume of the gas adsorbed is proportional to the partial pressure of the gas in compliance with the Freundlich's adsorption isotherm. Adsorption also depends on the active surface area of the adsorbent material. Various constituents of an air sample are absorbed in amounts inversely proportional to their volatility. Several solid adsorbents are widely used such as activated carbon (charcoal), silica gel, alumina, molecular sieves, porous polymer beads and gas chromatographic supports.

The method involves drawing a known volume of air at a controlled flow rate through a small tube packed with an appropriate sorbent. As the air passes through the tube, the molecules of the contaminant are adsorbed onto the surface of the sorbent and remain chemically and physically unchanged. The contaminant is then desorbed from the sorbent by a suitable solvent or thermally for subsequent analysis. Thermal desorption has the advantage that the entire sample can be removed from the sorbent and analysed at once, this increases the sensitivity of the analytical method.

Several factors are to be considered in selecting a suitable adsorbent. A high relative surface area is important to maintain a large contact area for adsorption and allowing maximum space between adsorbent granules for maximum air flow. Silica gel and alumina offer surface areas of 200-600 m^2 g^{-1}, activated carbon offers 500-2000m^2 g^{-1}, while gas chromatographic columns offer considerably large surface areas per g. The adsorbing properties (adsorption affinity for polar or non-polar compounds are also important for selecting adsorbent to collect a given gaseous contaminant. The polar adsorbents have more affinit for polar molecules. Polar adsorbents such as silica gel, can therefore be used either for short duration sampling in atmospheres that contain relatively high concentrations of contaminants or in atmospheres that are sufficientl low in moisture content so that the adsorbent does not become saturated with water vapor before the sampling is complete.

Silica gel is an amorphous form of silica formed by reacting sodium silicate with sulfuric acid. It is electrically polar and attracts polar molecules to the active sites on its surfaces. The use of silica gel offers the advantage that desorption of contaminants can be easily accomplished, it can effectively be used to collect certain inorganic substances for which charcoal is unsuitable. Charcoal is an amorphous form of carbon formed by burning wood, nutshells, animal bones and other carbonaceous materials. It is electrically non-polar and adsorbs organic vapors and gases in preference to atmospheric moisture.

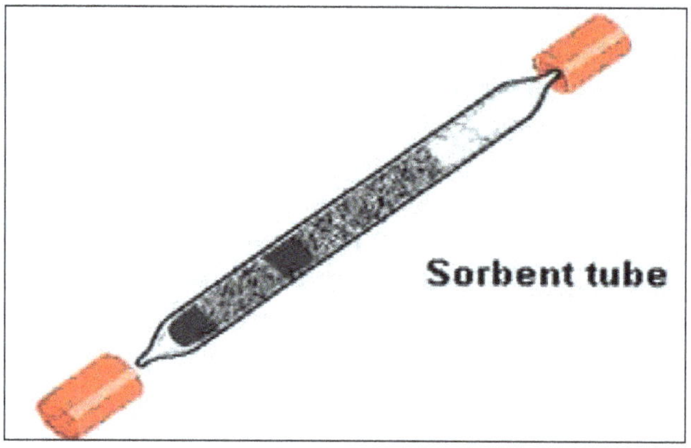

Adsorbent Tube.

Charcoal has an extensive surface area as large as 930 $m^2\ g^{-1}$. In industrial settings, organic solvent vapors are found as mixtures and not in a single pure form. Therefore, air sampling and analytical technique which is capable of collecting, separating, identifying and determining the concentrations of each individual constituent of a mixture is required. Gas chromatographic tubes filled with charcoal are preferred for sampling in such conditions. The organic vapors are adsorbed onto charcoal during sampling. The adsorbed material is desorbed by extracting in CS_2 and subsequently analysis is performed with flame ionization or electron capture gas chromatography. A disadvantage of the method is that desorption of the contaminant is not always 100%, and dilution of the sample occurs from solvent extraction resulting in lowered sensitivity and the solvent CS_2 is also toxic and inflammable. Further compounds such as O_3, NO_2, Cl_2, H_2S react with activated charcoal; hence these cannot be collected by this method. Amines and particularly the aromatic amines like methylaniline, N,N-dimethylaniline, p-nitroaniline, o-nitroaniline are not easily desorbed from charcoal and must be collected with other adsorbents like silica gel.

Adsorbent Tubes

Impregnated Solid Adsorbents

The development of Ion chromatography as an analytical tool for analysis of the ionic forms of gaseous agents led to the development of solid sorbent tubes impregnated with absorbing reagents for the collection and analysis of reactive gases. The sampling with impregnated solid adsorbents involves the collection of a gaseous contaminant in standard solid sorbent tubes impregnated with an appropriate absorbing reagent which changes the collected gas to a more stable ionic form. This is achieved by drawing the air through the sorbent tubes at a controlled flow rate for a known period of time. The sample is then desorbed with an appropriate solvent, followed by ion chromatography analysis.

Impregnated solid sorbent tubes containing have been utilized for SO_2 and formaldehyde with efficienc between 95 and 100%. Impregnated solid adsorbents, facilitates the collection and analysis of certain gaseous contaminants which cannot be analysed by the conventional methods. The sorbent tubes can easily be handled in the field without any kind of spillage

Passive Dosimeter

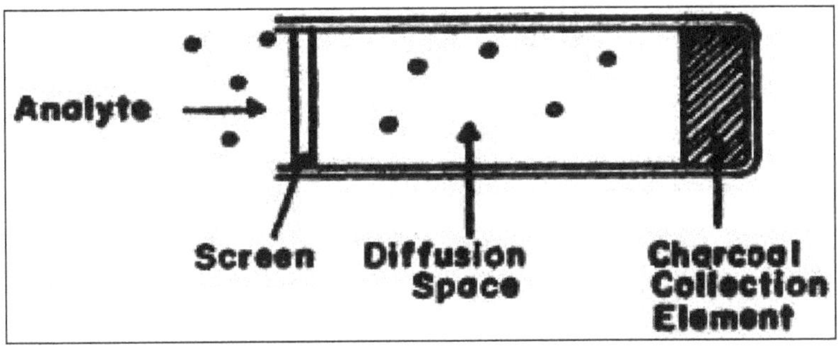

Passive dosimeter is a simple sampling device for sampling both organic and inorganic vapors, usually employed in industrial settings. This sampling technique does not require any sampling pump, has a low unit and capital cost. The technique is based on Fick's first law of diffusion or the permeation properties of the gas to be measured. The dosimeters consist of solid sorbent tubes and the samples can be analysed by gas chromatography. Collection of contaminant gases takes place by two different mechanisms. The gas is first adsorbed into a bed of uncoated solid adsorbent like charcoal or silica gel or the contaminant may react with a chemical coating present on the collection surface. Charcoal is generally used as sorbent for collection of non-polar organics while silica gel is used for polar gases and chemical coatings for reactive gases. The method offers almost 100% collection efficienc as the concentration of the contaminant gas at the boundaries (surface) is almost zero and a gradient of concentration exists in the stagnant air layer. Each gas has a unique diffusion coefficien in air, and due to the gradient of concentration, is transported to the surface with a velocity dependent on ambient concentration. Passive sampling has been used for measurement of ambient SO_2 with the lead peroxide candle method. In this method lead peroxide is exposed within a louvred box to ambient air over a month. The SO_2 reacts with lead peroxide, forming sulphate, which is measured by turbimetric methods using $BaSO_4$. A major limitation of the method arises from the much lower transference rates of pollutants to the absorption medium. Passive sampling can only be used for long-term sampling or for determination for high pollutant levels.

Condensation Method

The condensation method is used to collect gaseous contaminants in liquid or solid forms primarily for identification purposes or to sample gaseous contaminants which are difficul to collect by other methods such as sulphur trioxide. The condensation of gases from the atmosphere depends upon cooling the gas stream to temperatures below the boiling or freezing point of the gases to be collected. Samples are collected by drawing the air sample through a single or series of cold traps immersed in dry ice and acetone, liquid air or in liquid nitrogen refrigerant bath system. The collection traps are double walled, with the sampled air passing through the space between the walls. The temperatures used in the condensation device are selected to ensure condensation of the gas by lowering its vapor pressure below 1.33mb to prevent significant evaporation during sampling. Condensation methods have the advantage that the contaminant is preserved in its natural state without any chemical reaction. The method is particularly useful if the concentrations of a given contaminant are low and a highly concentrated sample is required for analysis. A major drawback of the method is that condensation units are bulky and is usually not portable to be applied in field measurements; secondly, unwanted gases and water vapor in the air may also condense out. This is prevented by employing sequential traps at progressively lower temperatures. The condensation systems cannot be employed for unattended operation.

Direct Reading Instruments

Direct reading instruments have the capability of accurately and immediately detect potentially hazardous concentrations of airborne contaminants. Direct reading instruments are of two types. The fir t type consists of those that produce a color change either in solution or detector (indicator) tubes through which the air sample is drawn or on chemically treated papers exposed to contaminated atmospheres. The second type comprises those that have electronic circuitry and are capable of sampling a volume of air, performing both qualitative and quantitative analysis internally and displaying the results on a dial, digital display, or strip chart recorder or printout.

Direct Reading Instruments with Indicators

Three types of direct reading colorimetric indicator systems are used for the determination of concentrations of gaseous contaminants: liquid reagents, chemically treated papers and detector tubes containing solid supports treated with chemical reagents. All the three systems utilize the chemical properties of contaminant gases to produce a reaction with a color-productive reagent. In instruments utilizing liquid reagents, the reagent solution is carried into the field with the air sampling unit such as impinge or bubbler. This method has been widely used to determine the concentrations of nitrogen oxides in air by passing a known volume of air at a controlled rate through Saltzman's reagent taken in fritted glass bubbler. The air is passed until a perceptible color change occurs. The concentration of nitrogen dioxide in air is inversely proportional to the time required to produce a perceptible color change which can be conveniently measured. A major disadvantage of this method in the field is that they are inconvenient and bulky to transport

Papers impregnated with chemical reagents were widely employed in the past to detect presence of toxic gases and vapors in work atmospheres; these are rarely used today because of the availability of more sensitive, sophisticated and accurate methods. Papers impregnated with mercuric bromide for the detection of arsine, lead acetate for detection of H2S, a mixture of o-toluidine and cupric acetate for the detection of HCN and detector tabs for CO. The observed time required for a color change after exposure of a specific paper or detector tab to an agent is an indication of the concentration present.

Detector tubes have been widely used as a convenient and economical tool for the detection of potentially toxic agents. They were first developed for monitoring of CO and H_2S. A detector tube unit comprises of a pump, a colorimetric indicator tube which is a sealed glass tube containing a granular material such as silica gel, alumina or pumice impregnated with a chemical reagent that reacts with the contaminant in the air stream as it is drawn through the tube. For sampling and measurement, the two sealed ends of the tube are broken and the specified end of the tube is inserted into the rubber septum of the inlet of the pump and affixe volume of air is drawn at a controlled rate. After a short specified time has been allowed for color development, the concentration is determined.

The concentration can be determined by either of the following three ways : 1) by comparing an absolute length of stain produced in the column of indicator gel or a ratio of the stain length to the total gel length against a calibration chart; 2) by comparing a progressive change in color intensity with a chart of color tints; or 3) noting the time required to produce an immediate color change in which the air volume sampled is intended to be inversely proportional to the concentration of the atmospheric contaminant. There are certain limitations of this method: 1) results obtained from matching tube color change with charts of tint is highly subjective among readers 2) results could be erroneous at high or low temperatures due to temperature dependence of the rate of chemical reaction; 3) lack of specificity of the indicator tubes, two or more gases can interfere with each other during the color producing reaction.

Electronic Direct Reading Instruments

Direct reading instruments have electronic sensors utilizing infrared and ultraviolet radiation, flame and photoionization and chemiluminescence capable of detecting and measuring airborne concentrations of gases and vapors in a few seconds. Most of these instruments are equipped with automatic continuous recording devices which generate real time data representing peak exposure concentrations at any point in time as well as time weighted data (averaging concentrations over time from a few seconds or minutes to full 8-hr work periods or longer. Most recently these direct reading instruments have been linked to microcomputers which allow immediate treatment and reduction of a mass of exposure data to a readily usable form on termination of sampling, correction of interferences, and automatic analysis for multi-component mixtures. The operating principles of direct reading instruments are based upon the physical and/or chemical properties of the gaseous agents they detect and quantify. Instruments based on physical properties comprise of an electronic detector or sensor that generates electrical signal in response to a physical phenomenon. For example the mercury vapor meter which is based on the absorption of UV light by Hg vapor showing a strong absorption line in the 253.7 mμ region of the UV spectrum. The instrument consists of an absorption chamber with a UV light source located at one end of the chamber and a photosensitive detector/sensor element located on the other.

Air is drawn through the instrument's absorption chamber where UV light is absorbed by Hg vapor present in the air stream. The presence of the Hg vapor reduces the UV radiation reaching the photosensitive detector element in proportion to the concentration present. The change in intensity of UV radiation reaching the photosensitive detector element which is connected to one arm of a Wheatstone bridge creates an unbalanced condition that is detected and displayed on a meter. In instruments based on the chemico-physical principle, the gas or vapor undergoes a chemical reaction and a physical method is used to detect the changes caused by this reaction. Either the consumption of one of the reactants or the production of one of the products is measured. Oxidation reduction reactions are examples of chemico-physical detection methods. The chemical part of the method is the oxidation–reduction reaction; the physical part is the measurement of electrons involved in the process. An example of such an instrument is the Mast Ozone meter in which O_3 is used to oxidize KI to molecular I_2 and KOH. The free I_2 produced reacts with a thin layer of H_2 that covers a wire electrode. The reaction consumes both H_2 and I_2 to yield HI. The removal of H_2 from the thin layer allows a polarization current to flow through the wire, regenerating the H_2 layer. For each O_3 molecule in the sample, two electrons flow in the circuit. The micro coloumb sensor counts these electrons and displays the concentration of O_3 in ppm.

Sampling of Water and Wastewater

A pre-requisite in developing a monitoring plan is to clearly identify the objectives of the monitoring. Water and wastewater monitoring may be undertaken to:

- Gaining an understanding of an aquatic ecosystem and the physical, chemical and biological processes that operate within it.
- Review of water quality within specified criteria

It is also important to consider the number of sites, number of replicates and the frequency of sample collection and amount of sample to be collected in planning a sampling design, to provide confidence in interpretation of results. A poor sampling design or too few samples can result in data that is too variable to reveal an impact, disturbance or trend. Samples should be collected from sites and times that provide a representative sample free from contamination and should not decompose, thus providing an accurate description of the overall quality of the water and wastewater. Sampling sites should also be located in safe and accessible areas, be well mixed to ensure a homogenous sample is collected and be easily identifiable for later sampling and comparable o er time.

Two types of sampling techniques are generally adopted for sampling of natural and waste water. These are grab or spot sampling and composite sampling.

Grab or Spot Samples

Grab samples are discrete samples that are taken at a location to provide a record of the water quality characteristics at that location and time of sampling. They do not provide any information about the concentrations outside that point in time. If grab samples are employed, a high number of samples are required to show the nature of change over time. However, taking manual grab samples is labour intensive and often impractical for long, intensive sampling plans.

Composite Samples

A composite sample is a sample consisting of two or more sub-samples mixed together in known proportions. Composite samples may be collected manually by combining grab samples, or by an automatic sampler. Compositing samples increase the temporal and spatial extent of sampling, without increasing the number of samples or sampling and analysis costs. These types of samples are used when the average water quality characteristics are of interest over a given period of time or volume of flow. They are more appropriate than grab samples when the distribution of constituents within the waste stream is random or when the variability within that stream is low. Composite samples are also useful when the determination of loads of constituents is required. However, compositing does have its limitations. A prior knowledge of the stream is required to determine if composite samples are appropriate, which is done by a pilot study of discrete grab samples. Additionally, compositing may mask variability within the waste stream by hiding peak and low concentrations. Composite samples are not appropriate for analytes that degrade during sampling or transport (e.g. dissolved oxygen, chlorine) or for easily contaminated samples such as microbiological sampling.

There are two basic types of composite samples in water sampling:

- Time-weighted samples.
- Flow-weighted samples.

Time-weighted samples are sub-samples of equal volume taken at constant intervals during the sampling period. For example, four samples are taken six hours apart to create a 24-hour composite sample. In flow-weighted sampling, the sub-samples are proportional to the effluen flow or volume during the sampling period. A flow-weighte sample can be created by taking samples at constant intervals but with varying sample volumes that are proportional to the flow at the sampling time; or by taking samples of equal volume that are taken at the time when fixed amounts of effluent have passed the sampling poin

Separate samples must be collected for chemical and biological analysis as the sampling and preservation techniques are different. Moreover for an accurate analysis, it is desirable to allow short time interval between sampling and analysis. For example temperature, pH and dissolved gases should be determined as quickly as possible preferably in the field. Similarly redox reactions can cause errors in analysis, microbial activity can reduce the phenol and COD values, change the NO_3^- — NO_2^- —NH_3 or alter the relative proportions of SO_4^{2-}, cause oxidation of S^{2-}, I^- and CN^-, reduce Cr(VI) to Cr(III) which precipitates readily, while Na, SiO_2 and B may be leached from the glass containers and metals may be adsorbed on the container walls. Colour, odour and turbidity may change with aging of the sample. Hence careful sampling and preservation techniques should be adopted. Table below lists the sampling, preservation and analytical methods along with the permissible values for potable water for some important water quality parameters.

Table: Summary of Sampling, Preservation and Analytical Methods and Permissible Values for Important Water Quality Parameters.

Parameter	Sampling Container	Preservation Method	Analytical Method	Permissible Value for Potable Water
pH	Plastic / Glass	Refrigeration	pH Meter	6.5-8.5
Colour	Plastic / Glass	Refrigeration	Visual/ Spectrophotometry	25 Hazen unit
Electrical Conductance	Plastic / Glass	None required	Conductivity Meter	2000 μScm^{-1}
Alkalinity	Plastic / Glass	Refrigeration	pH metric Titration	600mg/L
Turbidity	Plastic / Glass	Refrigeration	Nephelometry	10 NTU
Hardness	Plastic / Glass	Refrigeration	Titrimetric	600 mg/L
DO BOD	Plastic / Glass	Refrigerate/Store in dark	Winkler's Method	>6 mg/L 2 mg/L
COD	Plastic / Glass	Acidify and Refrigerate	Titrimetric	10 mg/L
NO_3^-, NO_2^-	Plastic / Glass	Filter and Freeze	Ion Selective / Colorimetric (UV)	20, 3 mg/L

NH_3	Plastic / Glass	Refrigerate	Ion Selective/ Colorimetric.	1.5 mg/L
Organic Nitrogen	Plastic / Glass	Refrigerate	Kjeldahl	100 mg/L as N
F^-	Plastic (teflon	Refrigerate	Ion Selective /SPADNS	1.5 mg/L
Cl^-	Plastic/Glass	Refrigerate	Argentometric	250mg/L
SO_4^{2-}	Plastic/Glass	Refrigerate	Turbidimetric	400 mg/L
CN^-	Amber glass	Refrigerate	Ion Selective/	0.05 mg/L
Cu	Acid washed Plastic/Glass	Acidify with nitric acid	AAS/ICP	1.5mg/L
Fe	Acid washed Plastic/Glass	Acidify with nitric acid	AAS/ICP	0.3 mg/L
Hg	Glass	Acidify with nitric acid	AAS/ICP (Cold Vapour)	0.001 mg/L
As, Cr (as Cr VI)	Amber glass	Acidify with nitric acid	Hydride	0.05 mg/L
Pb	Acid washed Plastic/Glass	Acidify with nitric acid	AAS/ICP	0.1 mg/L
Se, Cd	Acid washed plastic/ Glass	Acidify with nitric acid	Hydride generation/ AAS/ICP	0.01 mg/L
Anionic Detergents (as MBAS)	Plastic/Glass	Refrigerate	Colorimetry	0.05 mg/L
Hydrocarbons, Oil and Grease	Glass	Refrigerate	Gas Chromatography	Absent
Pesticides	Glass	Refrigerate	Gas Chromatography	Absent
Phenolic Compounds	Amber glass	Refrigerate	Gas Chromatography	Absent
Total Coliform (MPN/100mL)	Plastic/Glass	Refrigerate	MPN Method	10

3

Understanding Air Quality

Air quality is the degree of how polluted or how clean the air is. Air quality index is used for reporting daily air quality. There is a system to monitor various parameters of air such as the concentration of air pollutants, wind speed and direction, etc., which is known as an air quality monitoring system. This chapter has been carefully written to provide an easy understanding of the varied facets of air quality.

Air quality is a measure of how clean or polluted the air is. Monitoring air quality is important because polluted air can be bad for our health—and the health of the environment. Air quality is measured with the Air Quality Index, or AQI. The AQI works sort of like a thermometer that runs from 0 to 500 degrees. However, instead of showing changes in the temperature, the AQI is a way of showing changes in the amount of pollution in the air.

The air in our atmosphere is mostly made up of two gases that are essential for life on Earth: nitrogen and oxygen. However, the air also contains smaller amounts of many other gases and particles. AQI tracks five major air pollutants

- Ground level ozone
- Carbon monoxide
- Sulfur dioxide
- Nitrogen dioxide
- Airborne particles, or aerosols

Ground level ozone and airborne particles are the two air pollutants that pose the greatest risk to human health in the United States. They are also the two of the main ingredients in smog, a type of air pollution that reduces visibility.

Causes of Bad Air Quality

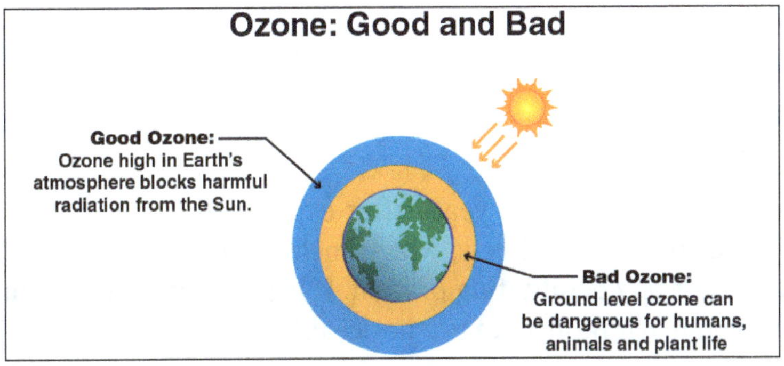

Ozone is a gas you've probably heard of as a layer high up in Earth's atmosphere. This ozone layer is a good thing—it helps block us from the Sun's harmful radiation. However, ground level ozone is bad for human health. It is created when sunlight reacts with certain chemical emissions (for example nitrogen dioxide, carbon monoxide and methane). These chemicals can come from industrial facilities, car exhaust, gasoline vapors and other sources.

Airborne particles are solid and liquid droplets suspended in the air. These particles become airborne at construction sites, smokestacks, car exhaust pipes, wildfires, volcanoes and many other places, too. The particles can also be formed from chemical reactions in the atmosphere.

Indoor Air Quality

Indoor Air Quality (IAQ) refers to the air quality within and around buildings and structures, especially as it relates to the health and comfort of building occupants. Understanding and controlling common pollutants indoors can help reduce your risk of indoor health concerns. Health effects from indoor air pollutants may be experienced soon after exposure or, possibly, years later.

Immediate Effects

Some health effects may show up shortly after a single exposure or repeated exposures to a pollutant. These include irritation of the eyes, nose, and throat, headaches, dizziness, and fatigue. Such immediate effects are usually short-term and treatable. Sometimes the treatment is simply eliminating the person's exposure to the source of the pollution, if it can be identified. Soon after exposure to some indoor air pollutants, symptoms of some diseases such as asthma may show up, be aggravated or worsened.

The likelihood of immediate reactions to indoor air pollutants depends on several factors including age and preexisting medical conditions. In some cases, whether a person reacts to a pollutant depends on individual sensitivity, which varies tremendously from person to person. Some people can become sensitized to biological or chemical pollutants after repeated or high level exposures.

Certain immediate effects are similar to those from colds or other viral diseases, so it is often difficult to determine if the symptoms are a result of exposure to indoor air pollution. For this reason, it is important to pay attention to the time and place symptoms occur. If the symptoms fade or go away when a person is away from the area, for example, an effort should be made to identify indoor air sources that may be possible causes. Some effect may be made worse by an inadequate supply of outdoor air coming indoors or from the heating, cooling or humidity conditions prevalent indoors.

Long-Term Effects

Other health effects may show up either years after exposure has occurred or only after long or repeated periods of exposure. These effects, which include some respiratory diseases, heart disease and cancer, can be severely debilitating or fatal. It is prudent to try to improve the indoor air quality in your home even if symptoms are not noticeable.

While pollutants commonly found in indoor air can cause many harmful effects, there is considerable uncertainty about what concentrations or periods of exposure are necessary to produce specific health problems. People also react very diff rently to exposure to indoor air pollutants. Further research is needed to better understand which health effects occur after exposure to the average pollutant concentrations found in homes and which occurs from the higher concentrations that occur for short periods of time.

Air Quality Index

Air quality index (AQI) is a numerical scale used for reporting day to day air quality with regard to human health and the environment. The daily results of the index are used to convey to the public an estimate of air pollution level. An increase in air quality index signifies increased air pollution and severe threats to human health.

In most cases, AQI indicates how clean or polluted the air in our surrounding is, and the associated health risks it might present. The AQI centers on the health effects that may be experienced within a few days or hours after breathing polluted air.

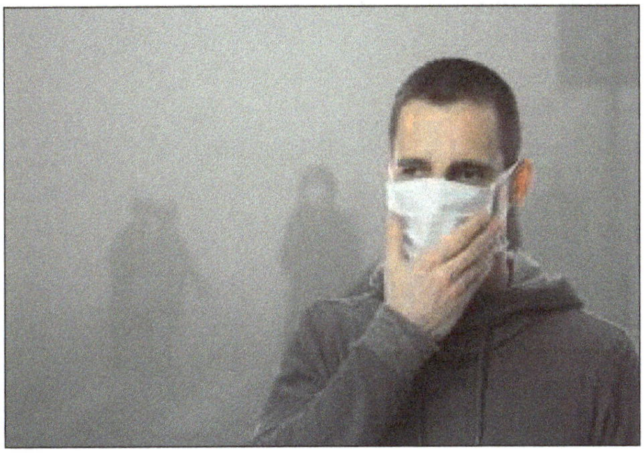

AQI calculations focus on major air pollutants, including particulate matter, ground-level ozone, sulfur dioxide (SO_2), nitrogen dioxide (NO_2), and carbon monoxide (CO). Particulate matter and ozone pollutants pose the highest risks to human health and the environment. For each of these air pollutant categories, differ nt countries have their own established air quality indices in relation to other nationally set air quality standards for public health protection.

Determination of Air Quality

On an hourly basis, the concentration of each pollutant in the air is measured and converted into a number running from zero upwards by using a standard index or scale. The calculated number for every pollutant is termed as a sub-index. The highest sub-index for any given hour is recorded as the AQI for that hour. In simple terms, AQI is like a yardstick that ranges from zero to five hundred (0-500).

The index is a relative scale, meaning, the lower the index, the better the quality of air and the lesser the health concern, and vice versa. The concentration of each pollutant varies; therefore, AQI values are grouped into ranges assigned to standardized public health warnings and color code.

For instance, an AQI value of 0 to 50 means good air quality with a limited possibility of affecting public and environmental health. On the other hand, an AQI value of 300 to 500 represents hazardous air quality with greater potential to affect public and environmental health. The commonly accepted value is 100, as it corresponds to the generally approved air quality standards as set to safeguard public health. AQI levels below 100 are highly satisfactory, while values beyond 100 are regarded to harm human health. As the AQI values get higher, it even poses more serious health concerns.

During days that the AQI is recorded to be elevated, the institution of public health might: offer advice to sensitive groups such as the children, those with respiratory problems, and the elderly to keep away from outdoor activities; take action to improve air quality by introducing measures for reducing emissions; or recommend the use of air pollution masks, especially in severe cases of air pollution.

Formula for Calculating Air Quality Index

Air quality index is calculated by a linear function by determining the concentration of the pollutant. The equation below is used to compute AQI.

$$I = \frac{I_{high} - I_{low}}{C_{high} - C_{low}}(C - C_{low}) + I_{low}$$

Where:

I = the (Air Quality) index

C = the pollutant concentration

$C_{\{low\}}$ = the concentration breakpoint that is ≤ C

$C_{\{high\}}$ = the concentration breakpoint that is ≥ C

$I_{\{low\}}$ = the index breakpoint corresponding to $C_{\{low\}}$

$I_{\{high\}}$ = the index breakpoint corresponding to $C_{\{high\}}$

Air Quality Index Categories

The AQI is divided in six categories and each category is meant to correspond to different health concern levels. Below is an explanation of the categories and their meanings:

- 0 – 50 Indicates "Good" AQI: At this level, the quality of air is deemed to be satisfactory, and air pollution poses little or no risk. This category has no health implications. Everyone can continue their outdoor activities normally.
- 51 – 100 Indicate "Moderate" AQI: This means acceptable Air quality. However, some pollutants might arouse modest health concern for a limited number of hypersensitive people. For instance, persons who are remarkably sensitive to ozone may experience respiratory

symptoms such as minor breathing difficulties Only very few hypersensitive people are recommended to reduce outdoor activities.

- 101 – 150 Indicate "Unhealthy for Sensitive Groups" AQI: This category may not be able to affect the general health of the public. Healthy people may experience slight irritations, and sensitive individuals will be slightly affected to a larger extent. However, children, older adults, and persons with lung disease are at greater risk from ozone exposure. Older children, adults and people with lung and heart disease are at greater risk from exposure to particulate matter and, therefore, should reduce sustained and high-intensity outdoor exercises.
- 151 – 200 Indicate "Unhealthy" AQI: In this category, every person might experience some adverse health effects. Sensitive group members may experience more serious effects. The hearts and respiratory systems of healthy people may be affected. Children, seniors and people with heart or lung disease should cut back sustained and high-intensity outdoor exercises or reschedule strenuous activities. General population should moderately reduce outdoor activities.
- 201 – 300 Indicate "Very Unhealthy" AQI: This would issue a health alert to mean that everybody may experience very serious health implications. Healthy people will commonly show symptoms. People with heart or lung disease will be significantly affected and will experience reduced endurance in activities.
- Older adults and children should stay indoors and avoid strenuous outdoor activities, significantly considering rescheduling or avoiding such activities. General population should reduce outdoor activities.
- Greater than 300 Indicate "Hazardous" AQI: Air quality at this level is life-threatening and would issue warnings of emergency conditions for the entire population. Healthy people will experience reduced endurance in activities and may also show noticeably strong symptoms. Other illnesses may be triggered in healthy people. Healthy individuals should avoid outdoor activities.

Elders and the sick should remain indoors and avoid outdoor exercise. It is recommended that children, seniors and the sick should stay indoors and avoid physical exertion. General population should avoid outdoor activities.

At a glance, the table below provides the Air Quality Index (AQI) categories:

Air Quality Index (AQI) Values	Levels of Health Concern
When the AQI is in this range:	Air quality conditions are:
0-50	Good
51-100	Moderate
101-150	Unhealthy for Sensitive Groups
151 to 200	Unhealthy
201 to 300	Very Unhealthy
301 to 500	Hazardous

Air Quality Monitoring

Atmospheric composition has been changing day by day with the addition of air pollutants which affects the biotic environment. The amount of air pollutant depends upon the source of air pollution and the ability of the atmosphere to absorb or disperses this emission. The concentration of air pollution varies spatially and temporarily, which causing the air pollution pattern to change with time and location. Presence of high amount of air pollutants adversely affects the health and prosperity of population. The major source of air pollutants include natural source and various industries such as power plants, incinerators and cement plants, domestic sources and vehicular pollution. To maintain the quality of air, policy maker at national and international level introduce certain rule and regulation.

1. Monitoring play an important part in controlling air pollution. It also provides information of various types of pollutant and particulate matter released out in the atmosphere. Monitoring therefore helps in the assessment, health and safety issue in the industrial plants. A good monitoring system requires equipment for sample collection, instrument calibration, data collection and their processing. There are various types of monitor are available which provide reliable and reproduc-ible data. Monitor is a device which measures or senses the physical or chemical properties of a substance. It generates an electrical signal and data acquisition systems (DAS) records these signal and makes a correlation to the concentration of pollutants.

Monitoring can be performed by two methods. One is stack and continuous emission monitoring system (CEMS). In case of CEMS, it is apply on any instrument for measuring the constituent of exhaust gas. CEMS equipment are quite expensive. Another method is parametric monitoring in which indirect measurement of the emission. In this method process or control parameter are correlate to check the levels of pollutant emission. It is commonly used methods for small emission source. It provides more flexible and less expensive means of monitoring

The selection of type of monitoring is depends upon various consideration like calibration or accuracy, maintenance requirements, chemical and physical properties of gas stream and pollutants, location, methods use in collection, processing and disposing of the sample, requirement based on quality control and quality assurance, safety management etc.

i. SO_x: Sulphur is an abundant element in the earth crust and found in the form of gypsum. The fuel like coal, oil, wood also contains small quantity of sulphur. SO_2 gas is produced by burning the material containing sulphur. First SO_2 is formed then it slowly oxidized to SO_3. Both SO_2 and SO_3 can form acid when hydrolyzed with water. There are two ways by which release of SO_2 is controlled i.e. dry and wet process. During operation stage we can control the production of SO_2 by using devices such as filters, wet scrubbers, condenser and by cyclone separators. Wet scrubbers are most commonly used method in controlling SO_X. In this a pressure drop is measured by manometer which indirectly measure SO_2. It is a less expensive method for monitoring of SO_2.The molecular weight of SO_2 is doubles that of elemental sulfur. So by monitoring the rate of fuel used, the emission of SO_2 is calculating by assuming complete combustion of fuel containing sulfur. Totalizers are used to monitored gaseous and liquid fuels. Another method of monitoring is by established a relationship between fuel used and steam production for a fuel.

ii. NO_x: Oxide of nitrogen is formed by material containing bound nitrogen or during the process of combustion when air combines with oxygen at a high temperature. Oxides of nitrogen are harmful when they are combining with hydrocarbon in the presence of sunlight forming harmful photo-chemical compound. Optimization of combustion is one of the control methods. Periodic testing of temperature, excess air and load can be monitored and established a correlation between these parameter and NO_x emission rate. An algorithm is developed to check the emission by DAS. Portable combustion analyser is used to monitor NO_x and excess air flow and temperature etc. NO_x can be measured by discrete analyser. NO_x is also measured by dilution extractive, cold/dry, direct extractive sample. It is measured in ambient air monitoring system.

iii. Hydrocarbon: In the atmospheric air hydrocarbons are complicated mixture of many substances like methane and non-methane.

2. Flue Gas Analysis: Combustion flue gas analysis has been used by power plant to optimizing the fuel/air ratio. The amount of excess oxygen and CO in the flue gas and least amount of NO_x and other greenhouse gases is measured by analyser. The stoichiometric point is where all fuel is reacted with the avail-able oxygen in the combustion air and no fuel or oxygen is left over.

i. Fuel: Fuel is those which on burning produce heat and power. Fuels used in various industry is available in three different forms i.e. solids (coal, wood, straw), liquid (diesel, crude oil) and gas-eous fuels(CO, H_2 and CH_4).The composition of fuel used is play an important role for optimization of combustion process.

ii. Flue gas: It is also known as stack gas generated through combustion processes. Composition of flue gas is depends on the type of fuel and condition during the combustion process. Flue gases are composed of high concentration of water and carbon dioxide, oxides of sulphur, oxides of nitrogen, fine dust, hydrogen halides, trace element such as mercury, nickel and dioxins etc. General compo-sition of flue gas is given below:

- Nitrogen (77%)
- CO_2, CO (13%)
- Oxide of N_2, H_2O, S and other (5%)
- Oxygen (5%)

All these compounds are unwanted by-product generated from various industries. Flue gas analysis indicates the air to fuel ratio. These flue gas components are part of air pollutants so these should be minimized or eliminated by cleaning procedure before released out in the atmosphere. These flue gases are vaporized or converted chemically into harmless gases with the help of cleaning procedure.

iii. Analyzer: Analyzer is a species–specific sensor. The function of analyzers is based upon various physical and physio-chemical principles like absorption, ionization, transmission and adsorption etc. Analyzers are classified into different categories:

- Portable analyzers: It can use to measured different location in very short period of time
- Fixed analyzers: It is used to measure in a fixed permanent location for long period of time.

- In situ analyzers: It is installed and worked in the process unit.
- Extractive analyzers: It is installed outside the process stream.
- Flue gas analyzer: It is an instrument which is capable of analysing the gases and measured the quantity present in the mixture or sample. There are various types of gas analyzer depending upon the principle used such as, gas chromatography, electrochemical analyzer, photometry (IR, Visible and UV), flame photometry, ionization, chemiluminescence, conductivity and paramagnetic etc.

3. Photometry: It uses the absorption of infrared (1000-10000 nm), visible (400-800 nm) and ultra-violet (200-400 nm) radiation by gases. Photometry based upon principle that gases absorb energy at certain wavelength and loss of radiation intensity is measured. It based upon Lambert-Beer law which states a linear relationship between absorbance and concentration of an absorbing species.

$$\log_{10}\frac{I_o}{I} = \varepsilon l c$$

Where,

I_0 -intensity of entering radiation

I-Intensity of emitting radiation

ε – extinction coefficie

C-concentration of gas

l-optical path length

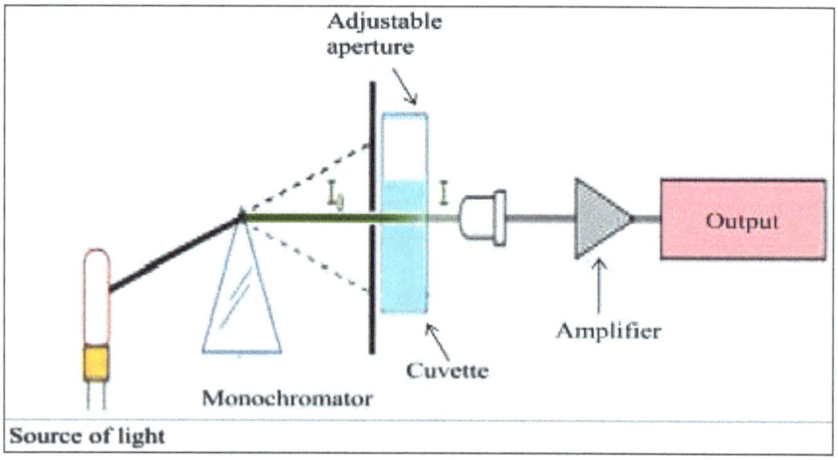

Basic instrumentation of Spectrophotometer.

4. Infrared absorption (IR) gas analyzer: Infrared radiation is absorbed by gases such as CO, CO_2, SO_2 or NO at specific wavelength particular to gas. The IR radiation from lamp is passed through the measuring cell filled with sample gas. Different molecules of gas absorb different frequencies of light. If concentration of gas is increased then IR absorption is increased and there is decrease in radiation intensity. A very low concentration of SO_2 and NO_2 is difficul to analyse because IR area is cross sensitive to H_2O.It is classified into two categories

i. Dispersive IR: It uses the radiation which is dispersed by prism. Two wavelength are used one is reference and other is that is being absorbs by sample gas. Gas concentration is measure by calculating the ratio of two wavelengths.

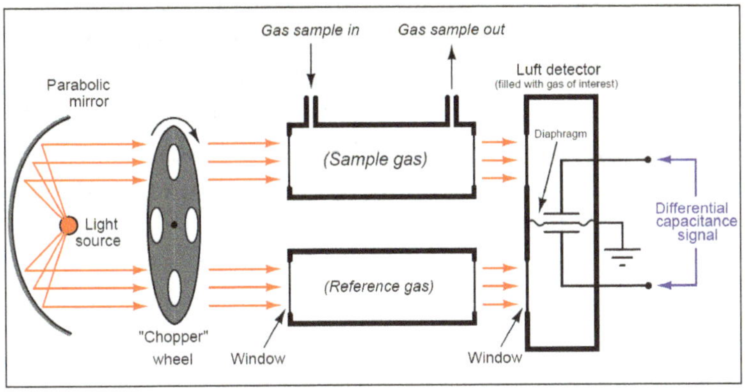

A Typical Infrared photometer.

ii. Non Dispersive IR: It uses a broad band of radiation from the lamp without dispersion. When components present in sample gas absorb IR, the intensity of radiation is used at specific spectrum when reached to the detectors. Two methods are used to detect the level of absorption. One is a gas detector and other is solid state IR detectors. NDIR instruments is primarily used for measurement of CO, NO, SO_2, H_2O, CH_4, CO_2 and many hydrocarbons.

Ultraviolet absorption: Gases like NO and SO_2 absorb UV radiation. The major limitation in using UV radiation is its high cost.

5. Fourier transform IR spectroscopy: FTIR spectroscopy is used to measured infrared active gases like SO_2, CO_2, CO, NO, HCl etc. It is based on the Michelson interferometer which has function of monochromator. In this radiation hits a beam splitter which reflects fifty percentage of the radiation and fifty percentages are transmitted. These two beam hits two mirrors which is perpendicular to one another and reflected back to the beam splitter. These reflected beams are recombines by beam spillter. The recombined beams are then passed through cell full of the product to be measured and focused on IR detectors.

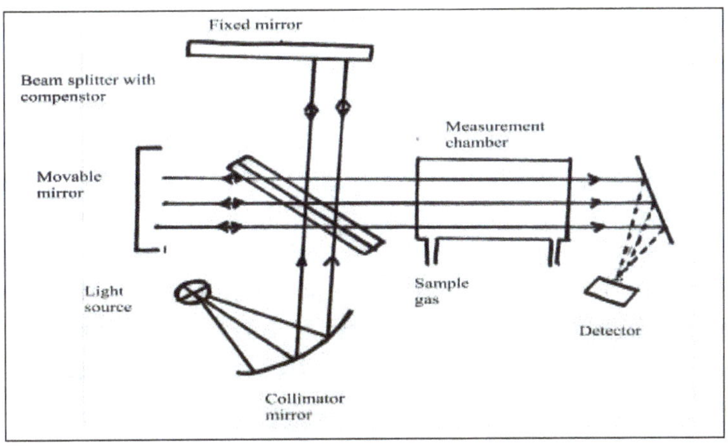

FTIR Spectrometer arrangement.

6. Flame ionization detection method: It is widely used to detect organic material in gas stream. In FID sample gas containing hydrocarbons is introduced into the flame (air or Hydrogen)

where hydrocarbon is ionized because organics compound are easily ionizable in hydrogen flame. The ions formed are extracted in an ionization chamber by applying an electric field using electrode, generating a current which is proportional to the concentration of hydrocarbon. The data evaluation is done by comparing the different sensitivity of different hydrocarbon compounds. In this process pure hydrogen gas (taken from pressurized gas cylinder or from electrolytic hydrogen generator unit) is flow through nozzle into the combustion chamber of flame ionization detectors. Combustion air from the atmosphere is passed through annular slit around the nozzle. Steady hydrogen flame produce by ignition produce very small ion density in the absence of organic compounds. Compared the value with sample containing organic compounds is detected. The temperature and pressure of the sample gas must be constant. FID provides a non-selective total measurement of organically bound carbon.

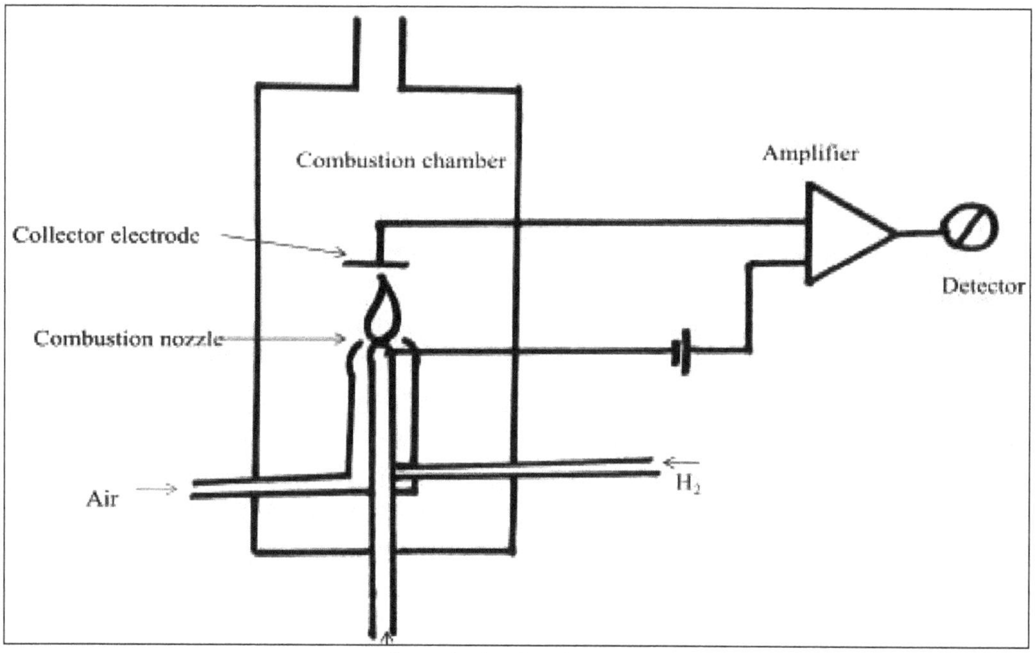

Flame ionization detector.

7. Chemiluminescence Method: It is used in determination of low concentration of oxides of nitrogen. It is based upon the principle that when NO is react with ozone then resulting in the generation of characteristics light radiation. The analyzers are consisting of ozone generator, a reaction chamber, an enrichment and photomultipliers detector. The oxygen present in the air is partially converted into ozone by UV radiation and electrical discharges. Then a constant flow of sample gas is passed through reaction chamber via another nozzle and both are mixed in. the chemiluminescence is optically filter and measured by photo multiplier. Reaction chamber is maintained at controlled pressure and temperature. To determine the concentration of NO_2, the sample gas is firstly passed through thermo-catalytic converter which reduces NO_2 to NO.

$$NO + O_3 \rightarrow NO_2^* + O_2$$

$$NO_2^* \rightarrow NO_2 + hv$$

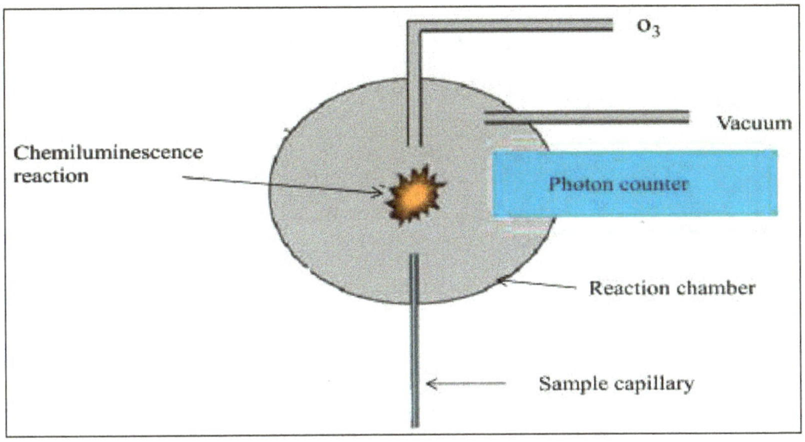

Showing working of Chemiluminescence.

8. Gas chromatography: In gas chromatography a sample is injected into the head of chromatographic column (contain liquid stationary phase adsorbed on the surface of inert solid) which is transported by inert gaseous mobile phase. Most commonly used carrier gases are nitrogen, carbon dioxide, helium and argon depending upon the type of detectors used. The temperature of the column must be controlled and maintained depending upon the boiling point of the sample. There are various types of detectors are used in gas chromatography for example ECD(electron capture detector) used for the monitoring of halides, nitriles, nitrate, peroxides and TCD(Thermal conductivity detector) is used universally for all. Finally result are recorded and displayed in data system attached with column.

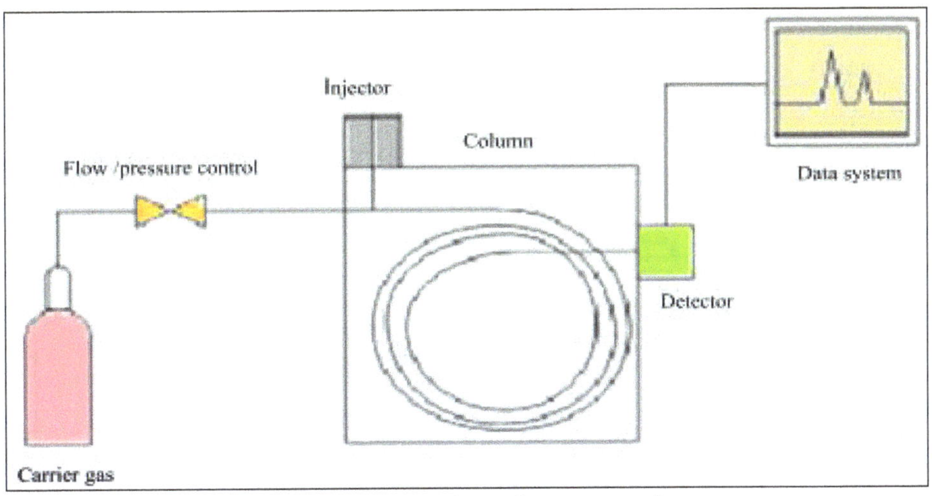

Showing working of Gas chromatography.

9. Thermal conductivity gas analyzer: Thermal conductivity is defined as ability to conduct heat at a specific rate. Each gas has different thermal conductivity for example nitrogen have 5.680 and sulphur dioxide has 1.950 thermal conductivity. The Zero and sample gas is passed onto the heated metal filaments. The amount of heat carries away by gases changes the rate of cooling of filament wire. Change in temperature resulting in change in resistance. The change in resistance is converted into electric current and an output signal. Thermal conductivity analyzer helps in analyzing mixture of two gases and in production of high purify gas. Other application includes food packaging mixtures, welding shield, leak detection mixture etc.

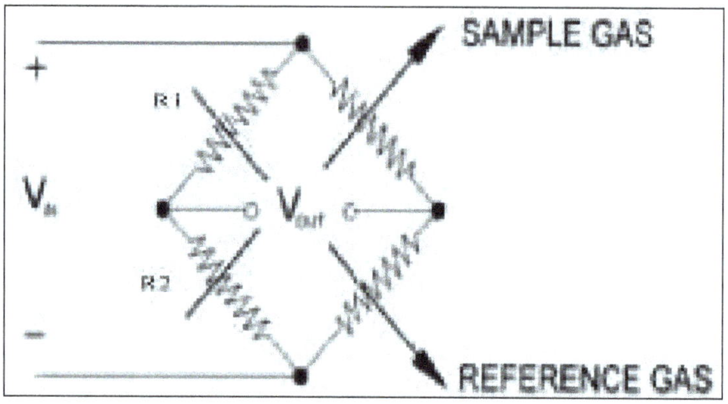

Thermal conductivity Detector Bridge.

10. Para magnetic gas analyzer: It is based upon the larger magnetic susceptibility of oxygen as compared to coexisting gas. This analyzer is best for measuring oxygen concentration in flammable gas. There is no need of using reference gas so save the cost. In this instrumental arrangement two glass spheres are filled with nitrogen gas are suspended with strong metal. These sphere are first kept in homogenous magnetic filled. Oxygen molecule in sphere having large magnetic susceptibility flow, the molecules are pulled toward strong magnetic field and sphere is moves away.

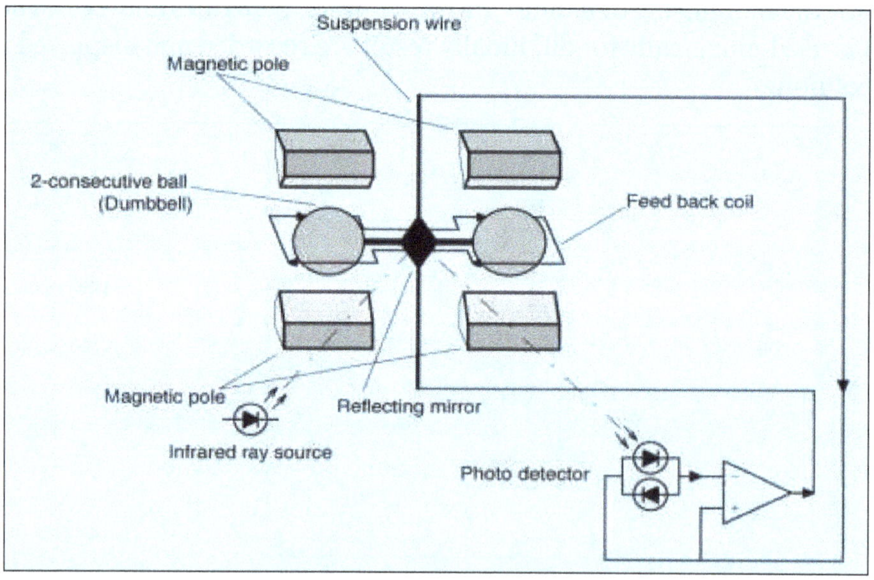

Paramagnetic gas analyser.

This deviation in the sphere is detected by light source, reflecting mirror and light receiving element. Current is then flowed through the feedback loop to control and sphere can return back to initial balanced state. This current used is proportional to the oxygen concentration.

11. Remote sensing monitoring: Remote sensing device uses the detector which measure the optical properties of the sample gas, simply by reflected and transmitted signal received after pathlength through the air. It found an enormous application in chemical processing, air quality monitoring, power generation and in transport by monitoring the gaseous emission.

- Light detection and ranging (LIDAR): It is used for online monitoring. It is use to track the pollutant for long period of time. The technique used is long path absorption, in which beam of laser light is reflected from distant retro reflector and come back to detecto

- Differential absorption lidar (DIAL): Pulses of two wavelength are directed into the air by tunable laser and a cooled, detector measured the backscattered signal form particles, gas molecules and from air molecules. Difference between the two can used to calculate absorption due to gases.

12. Less common measurement methods: Colorimetric, heat change, conductometry and potentiometry are measurement methods used for continuous emission monitoring of gas stream. Colorimetric method, the sample gas is brought in contact with suitable reagent and change in color is measured by photometric basis. In conduct metric method, the sample gas is introduced into suitable liquid reagent and change in conductivity is measured after completion of reaction. In heat change method, the temperature increase given of during exothermic catalytic oxidation of combustible gas component is measured. In potentiometric method, the sample gas in introduced into buffered electrolyte solution and change in ion concentration is measured by ion sensitive electrode change.

13. Flue Gas Analyzer by Companies: Now a day many companies are involving in manufacturing of flue gas analyzer. Some of them are given below:

- The Zirconium oxide fuel cell oxygen analyzer (Z_rO_2) has been used for measuring combustion flue gases. First used in 1970 in power generation industry but now used for many combustion process. All automobiles are now using these sensors for controlling fuel-air ratios. The main advantage of using these technologies is that its operation at hot combustion flue gases, it can easily used because theses sensors heated at 700-7500C. The sensors can directly place into the flue gas stream on the end of probe and there is no need of sampling system. The sensors are robust and can withstand the sulfur components found in many fuels. These sensors can be calibrated in place and on line. Automated calibration is also available.
- NOVA flue gas analyzer: NOVA 7200 flue gas analyzer system using various technologies for different gases. For example carbon monoxide and carbon dioxide by infrared detectors and nitrogen dioxide, CO_2, O_2 by electrochemical sensors. It is accurate and durable method of monitoring.
- Testo 340 flue gas analyzer: It is used to measure CO, NO, NO_2 and SO_2. It is easy to use. It has precalibrated gas sensors for quick and easy sensor change.
- Wohler A 550 industrial flue gas analyzer: It can measure NO_x and SO_x emission with 0.1 ppm resolution. It is used in complete analysis and inspection of burners and boilers.

Air Quality Modeling

Air quality modelling comprises two very distinct approaches,. The first approach is known as physical modelling, and attempts to reproduce observed meteorological and air quality conditions through subjecting scaled replicas of a domain to a series of controlled flow regimes with varying physical input parameters, e.g. air temperature, velocity, etc. Modelling of this nature is typically performed in a wind or water tunnel. A scale model of a specific structure, i.e. building, street, terrain, etc., is placed in the tunnel and subjected to particular flow conditions. Tracer elements representing airborne pollutants are then released upwind/stream of the structure and its transport downwind/stream monitored using appropriate sensing instrumentation. The second approach is classified as mathematical or computational modelling, and attempts to reproduce or predict air quality scenarios, by the intimation of mathematical and physical relationships. When these relationships become too tedious or complex to be used analytically, they are often expressed in algorithmic form and solved using computers.

Numerical modelling of air quality scenarios has been identified as being important in several ways. These are:

- To aid in the evaluation of source–receptor relationships so that responsibility for specific impacts can be apportioned.
- To aid in project planning, site evaluation and/or environmental impact of present/future sources.
- To enable the evaluation of existing sources in relation to compliance with legislation.
- To permit the evaluation of proposed abatement and control strategies, in relation to short and/or long term issues.
- To permit the assessment of episodic tactics and disaster aversion strategies.
- To optimize emission inventories and operating conditions while ensuring compliance with legislative controls.
- To forecast in real time concentrations of accidental releases.

Wyngaard established what has now become a widely accepted view, that a complete treatment of the physical dynamics of the atmosphere is computationally impossible and will be in the foreseeable future. This conclusion was based on two factors; the inability of current computational technology to provide a complete solution; and a general lack of comprehension regarding the complexities of atmospheric systems and related phenomena (i.e. deposition, precipitation, etc.). Mathematical modeling presents itself within this argument as a half-way house, often affecting a partial treatment which can be accurate enough to satisfy the empirical requirements of a given modelling application. For this reason, mathematical modelling is viewed as one of the best mediums/tools through which scientists and environmental engineers can investigate the dynamics of the atmosphere.

Historically two equally valid mathematical approaches have been employed in describing air quality dynamics. These are:

- The empirical/statistical approach attempts to express the entire spectrum of system behavior using statistical distributions and probability theory.
- The deterministic approach seeks to employ physical laws and simplifying assumptions, to construct a dynamic model based on partial differential equations

Of the two approaches, deterministic modelling is commonly favoured over a statistical approach as it deals with general case solutions and can be easily applied to decision making processes.

Simple Air Quality Models

Box Models

The simplest approach to estimating pollutant concentrations over a given domain, is to implement a single box model. The model likens the airshed of a domain to a rectangular box, inside which the mass of the pollutant is fully conserved. The box is orientated in such a way that the directional component of the wind velocity is both perpendicular and incident to one face of the box. The model also assumes that the incoming pollution is instantaneously mixed with the surrounding air, creating a homogeneous concentration throughout the airshed. The mass conservation constraint permits the construction of a mass balance equation of the form:

$$\frac{dCV}{dt} = QA + uC_{in}WH - uCWH$$

where,

V is the volume described by the box

C is the homogeneous species concentration within the airshed

C_{in} is the species concentration entering the airshed

Q is the emission rate per unit area of sources within the box

u is the average wind speed normal to the box

A is the horizontal area of the box (L × W)

W is the width of the box and H is the mixing depth.

Integrating equation $\frac{dCV}{dt} = QA + uC_{in}WH - uCWH$ provides a steady state estimation of species concentration, assuming the dynamics of the mixing depth to be quasi-stationary and the source emissions to be constant. In the case of reactive species, the chemical reaction dynamics can be incorporated into the mass balance equation as could wet and dry deposition effects

Work presented by Jensen & Petersen, demonstrated a good agreement between observed daily average pollution levels and predictions generated by a single box model. The concentration forecast was shown to lag the observed data by approximately 2 hours in predicting the transient behavior of the pollutant across the domain. If a best fit time-shift correction is applied to the pollutant concentration forecast, then the magnitude of the predicted pollutant concentrations lies within a 15% error margin with respect to the observed data trend.

Topcu et al. employed a single box model in estimating the daily polluting concentration over the Erzurum city in Turkey. Regression coefficient ranging from 87.5% to 97% were obtained when the predicted levels of sulphur dioxide were compared with the observed daily averages. Such high regression factors are typically rare in single box modelling applications though they do serve to demonstrate the possible accuracy of the approach.

The application of the single box model is confined to those problems whose objective is the estimation of average pollution levels, across a specific airshed. The size of the airshed is typically large, given that the founding assumptions of homogeneity and instantaneous mixing break down at smaller scales of interest. It is worth noting that by allowing the mixing depth (*H*) to be a function of the downwind distance *x*, so-called box models can be employed over smaller scales. In addition, the single box model is incapable of imparting any spatial information regarding the dispersive nature of a pollutant. This precludes the box model approach from a significant proportion of air quality modelling applications. The method is computationally fast and is capable of providing satisfactory predictions, particularly for scenarios where detailed information on the domain and meteorological conditions is unavailable.

Gaussian Model

The Gaussian model forms the basis for the majority of air pollution models, and is the most well-known and documented approach. The model presupposes that the dispersion associated with the polluting species can be described by a modified Gaussian or 'normal' distribution curve. A three-dimensional axis system is employed to provide a downwind, crosswind and vertical resolution. The species concentration is defined as being proportional to the emission rate of the source, diluted by the wind velocity at the source of emission. The dispersion behavior of a pollutant is determined by the standard deviations associated with the Gaussian distribution function. These standard deviations are typically functions of atmospheric stability, localized turbulence and distance downwind from the source. The model is usually aligned so that the downwind axis corresponds to the direction of the prevailing wind.

The model equation is derived from basic considerations of the diffusion of gaseous matter in three-dimensional space:

$$C = \frac{Q}{2\pi u \sigma_y \sigma_z}\left(\exp\left(\frac{-y^2}{2\sigma_y^2}\right)\right)\left(\exp\left(\frac{(-h-z)^2}{2\sigma_z^2}\right) + \gamma_G \exp\left(\frac{-(h+z)^2}{2\sigma_z^2}\right)\right)$$

where,

C is the species concentration at a location (x, y, z)

Q is the source emission rate

u is the average wind speed normal to the box

σ_y is the standard deviation of the horizontal crosswind distribution of the plume concentration and is a function of the downwind distance x

σ_z is the standard deviation of the vertical crosswind distribution of the plume concentration and is a function of the downwind distance x

h is the effective source height to which the plume has rise

r_G is the ground reflection coefficient wh $(0 \le r_G \le 1)$

y is the crosswind distance and z is the receptor height above ground.

The effective source height (h) or plume rise is the height to which an emission will initially rise as a result of thermal buoyancy and vertical momentum. The upward movement of the plume is retarded on mixing with ambient air reaching an equilibrium point when the internal energy of the plume is equal to that of the surrounding atmosphere.

Several assumptions are implied in the derivation of equation $C = \frac{Q}{2\pi u \sigma_y \sigma_z} \left(\exp\left(\frac{-y^2}{2\sigma_y^2}\right)\right)\left(\exp\left(\frac{(-h-z)^2}{2\sigma_z^2}\right) + \gamma_G \exp\left(\frac{-(h+z)^2}{2\sigma_z^2}\right)\right)$, including:

- The emission characteristics of the source are uniform and time invariant.
- The meteorological conditions within the domain of interest are homogeneous and time invariant.
- The topography of the domain is weak (i.e. flat), so as not to affect pollutant dispersio
- The plume is symmetrical with a straight line trajectory.
- The emission originates from a point source where the pollutants are both conservative and passive.
- A proportion of the plume is reflected back from the ground plane
- No chemical reactions or deposition effects are accounted for

These assumptions are not borne out in reality, as emissions and meteorological conditions can be highly variable and non-linear by nature, i.e. a doubling in emissions may not incur a corresponding change in the downwind concentration of pollutants, or for that matter in the concentration of any secondary pollutants. This factor is particularly true in conditions of increasing turbulence, in which distortions in the trajectory and dispersion of the plume can mask out any discernible relationship between the downwind pollutant concentration and its source. Pollutants are also seldom passive and emissions are typically dependent on plant operation. In light of such limitations, the Gaussian model can only be considered workable when such factors are static enough to be regarded as homogeneous. The predictive capabilities of the model are documented by the Canadian Standards Association, CSA.

For distances less than about 1 km under ideal conditions of uniform flat terrain, steady meteorology etc., the prediction of the maximum downwind ground level time integrated concentration is only accurate to within 20% for a ground level release and to within 40% for an elevated release. The accuracy of the model within $\pm 2\sigma_y$ off the centerline is estimated to be within a factor of two. When the Gaussian model is applied to reasonable conditions of meteorology, terrain, and distances up to 10 km, the scatter of the estimate of the average time integrated concentration is expected to increase and the accuracy is then estimated to be within a factor of two.

Under certain conditions the equation $C=\frac{Q}{2\pi u\sigma_y\sigma_z}\left(\exp\left(\frac{-y^2}{2\sigma_y^2}\right)\right)\left(\exp\left(\frac{(-h-z)^2}{2\sigma_z^2}\right)+\gamma_G\exp\left(\frac{-(h+z)^2}{2\sigma_z^2}\right)\right)$ can be reduced to a variety of simplified forms. The behavior of the Gaussian model depends heavily upon the correct calculation of the dispersion coefficients σ_y and σ_z. The most common method of calculation is the semi-empirical method.

Semi-Empirical Calculations of σ_y and σ_z

In the likelihood that specific information regarding the variation of the wind is unattainable, a second approach, utilizing routine meteorological observations to provide a semi-empirical estimation of the dispersion coefficients can be considered. Work by Smith and Singer & Smith led to a method for estimating the stability of the atmosphere using a wind direction trace obtained over 1 h. Five stability classes were proposed from the empirical data collected. The dispersion coefficient were calculated using equation $\sigma_y = ax^b \quad \sigma_z = cx^d$, where the relevant coefficients referenced to the appropriate stability class, can be seen in Table. Note both σ_y and σ_z have units of m where x is in km.

$$\sigma_y = ax^b \quad \sigma_z = cx^d$$

A similar approach was suggested by Pasquill. Pasquill advocated the use of fluctuation measurements for dispersion estimates but provided a scheme 'for use in the likely absence of special measurements of wind structure'. The scheme did not require sophisticated measurement equipment but utilised basic values of wind speed, insolation and cloudiness in the estimation of atmospheric stability. However, it was Gifford & Sklarew who restated the Pasquill stability classes in terms of σ_y and σ_z so as to permit their use in the Gaussian plume equations. A comparison between the gustiness classes and the Pasquill stability classes are given in Gifford. Turner and Gifford graphically related the Pasquill dispersion parameters to the downwind distance from the source. The dispersion coefficien is directly obtained by estimation from the graph, where the curve represents the value of the dispersion coefficient at a specific stability cla

A formula method of estimating the dispersion coefficien was proposed by Martin. Again, an exponent law was employed in obtaining the coeffici t values for downwind distances, over short and long scales of interest.

$$\sigma_y = ax^{0.894} \quad \sigma_z = cx^d + f$$

Similar formulations have been proposed by several workers. The latter is of particular note as estimates of the Pasquill dispersion coefficients over urban and rural domains are provide

Advanced Modifications and Applications

In order to overcome several of the limitations inherent within 'simple' Gaussian models, many workers have suggested additional modifications. Various analytical extensions of the Gaussian Plume Model, specifically the extension of the model into polar co-ordinate space so that an analytical solution could be developed, utilising wind velocity input data in a wind rose format. This approach sought to overcome the dependence of the Gaussian model on a constant velocity input vector, and to provide a facility for a 360° scope of prediction.

A segmented approach was developed to account for variations in the source emission and meteorology. The virtual plume is dissected into a number of segments whose dynamics are functions of the emission rate and the prevailing meteorological conditions, and are oriented according to the time varying wind direction. The model suffers from supposition effects, where the error associated with a pollutant concentration close to the ends of the segment, varies proportionally with the angle between the consecutive segments.

Table. Coefficients and exponents for gustiness stability classe

Stability class	a	b	c	d	Classificatio
A	--	--	--	--	Highly unstable, η >> 90°
B1	0.36	0.86	0.33	0.86	Unstable, 45° < η < 90°
B2	0.40	0.91	0.41	0.91	Unstable, 15° < η 45°
C	0.32	0.78	0.22	0.78	Neutral, $\eta \approx$ 15°. Core of trace is unbroken and remains in the sector for 1 h.
D	0.31	0.71	0.06	0.71	Highly stable η < 15°. Trace approximates a straight line.

η = fluctuati n in wind direction.

Table. Coefficients and exponents for pasquill stability classe

	$x \leq 1$ km				$x \geq 1$ km		
Stability class	a	c	d	f	c	d	f
A	213	440.8	1.941	9.27	459.7	2.094	−9.6
B	156	106.6	1.149	3.3	108.2	1.098	2.0
C	104	61.0	0.911	0	61.0	0.911	0
D	68	33.2	0.725	−1.7	44.5	0.516	−13.0
E	50.5	22.8	0.678	−1.3	55.4	0.305	−34.0
F	34	14.35	0.740	−.035	62.6	0.180	−48.6

More recently, van Jaarsveld *et al.* proposed a Gaussian 'operational model for priority substances' to predict the concentration and deposition removal of low-reactive pollutants. Comparison between the long range transport predictions and the data obtained from the LML-National Monitoring Network in the Netherlands show good correlation and demonstrated that the OPS model is capable of accurately describing 'the meteorological variability' and the 'influence of sources at large distances', i.e. under averaged conditions.

The results from the AERAD model (Atmospheric Dispersion Model for Emergency Response; Zhuang et al., 1993) demonstrated that the Gaussian approach can be used to 'approximate the real life situation' of pollutant transport.

Despite the efforts of workers, the limitations of the Gaussian model precludes its use in cases where the short term prediction of species concentration (i.e. sub-hourly averaged values), or the prediction of species concentrations relative to complex environmental constraints, are the primary objective.

Eulerian Models

The Eulerian approach to dispersion modelling solves the conservation of mass equation for a given pollutant species of concentration c. A stationary or 'normal' frame of reference is assumed, with the dispersion phenomena being calculated as a concentration 'field' relative to the domain. The physical conditions within the reference domain are generally regarded as being turbulent so that any dependant variable is composed of an 'average' component, denoted by an overbar, and a 'fluctuating' component, denoted by a prime, i.e. $U = \overline{U} + U'$ here U is the Eulerian wind field vector $U_{(x,y,z)}$. Given that $\langle \overline{U} \rangle = \overline{U}$ and $\langle \overline{U}' \rangle = 0$, where $\langle \; \rangle$ represents the ensemble average or theoretical mean and that $c = \langle c \rangle + c'$, then the general form for the equation of conservation of pollutant species c is:

$$\frac{\partial \langle c_i \rangle}{\partial t} = -\overline{U} \cdot \nabla \langle c_i \rangle - \nabla \cdot \langle c_i' U' \rangle + D\nabla^2 \langle c_i \rangle + \langle S_i \rangle$$

where,

c_i is the concentration associated with the *i*th species

D is the molecular diffusivity an

S_i is the sink/source term of the *i*th species and accounts for chemical reactions, deposition and emission sources

The terms on the right hand side of equation above represent:

- Rate of advection of pollutants.
- Rate of turbulent diffusion of pollutants
- Rate of molecular diffusion of pollutants
- Contribution of any sources or sinks.

The term $\langle c_i' U' \rangle$ represents the turbulent atmospheric diffusion eddies whose magnitude and effect are significantly greater than that of molecular diffusion. For the majority of cases it is usual to ignore the molecular diffusivity term as its overall contribution will be negligible. The eddy diffusivity term is unresolvable as c' is unknown and a suitable description of U' is more often than not impossible to obtain. Consequently, the diffusivit term, $\langle c_i' U' \rangle$, has to be modelled if equation $\frac{\partial \langle c_i \rangle}{\partial t} = -\overline{U} \cdot \nabla \langle c_i \rangle - \nabla \cdot \langle c_i' U' \rangle + D\nabla^2 \langle c_i \rangle + \langle S_i \rangle$ is to be closed. A variety of closure methods currently exist.

The 'zero order' method is the simplest of methods and assumes the rate of diffusion to be constant. Obviously this assumption is invalid for the majority of atmospheric applications. The gradient-transport method of closure, or more formally, the 'first order' closure method, forms the basis of the majority of atmospheric transport models, and assumes that:

$$\left\langle c_i' U' \right\rangle = -K\nabla\left\langle c_i \right\rangle$$

where K is the eddy diffusivity tensor

Equation $\left\langle c_i' U' \right\rangle = -K\nabla\left\langle c_i \right\rangle$ assumes that the pollutant species diffuses at a rate that is proportional to the localised concentration gradient. The eddy diffusivity, K, accounts for the increased rate of dispersion due to localised turbulence and atmospheric instability. K is typically a function of localised meteorology, height (z) and surface characteristics. The errors introduced by 'first order' closure depend on the behaviour and accuracy of the eddy diffusivity term K. In three dimensions, K becomes a 3 × 3 tensor. A general simplification is to assume that the primary mode of diffusive transport lies along the vector of the turbulent eddy, so that the eddy diffusivity tensor is diagonal and cross vector diffusivities are negligible, that is

$$K = \begin{bmatrix} K_{xx} & 0 & 0 \\ 0 & K_{yy} & 0 \\ 0 & 0 & K_{zz} \end{bmatrix}$$

A horizontal diffusivity is defined, assuming $K_{xx} = K_{yy} = K_H$, and is further simplified if the model is oriented in 'sync' with the mean flow so that the contribution made by horizontal diffusivity in the x direction can be assumed to be negligible in comparison with that imparted by advection. It is acknowledged that in the majority of flow scenarios, the primary cause of diffusion emanates from the vertical movement of air resulting from variations in temperature and pressure, as well as crosswind dispersion due to wind meandering and turbulence.

The conditions of validity for first order closure were stated by Lamb, who concluded that K theory was only applicable when:

$$\frac{t_e}{T_c} << 1$$

where t_e is the maximum time over which the turbulent eddy maintains its integrity and T_c is the time scale associated with the mean flow field. The changes in the mean concentration field are required to have a larger scale than the time scale associated with the turbulent mean flow field. This is often violated near a strong isolated source in unstable conditions.

Higher order methods of closure have been proposed by several workers, though these methods exhibit a tendency to be highly complex, computationally expensive and experimental. Consequently, second order methods are not as widely applied in comparison to more simple techniques. Zannetti concluded, on second order closure, that the practical applicability of higher order closure models is still questionable, even though recent results are encouraging.

After selection of a suitable closure method, equation $\frac{\partial\langle c_i\rangle}{\partial t}=-\overline{U}\cdot\nabla\langle c_i\rangle-\nabla\cdot\langle c_i'U'\rangle+D\nabla^2\langle c_i\rangle+\langle S_i\rangle$ is typically solved in one, two or three dimensions on a discrete mesh architecture, using a suitable numerical method, e.g. fin te difference, finite volume, finite element. Some workers have obtained encouraging results from the steady state case. A comparison of numerical methods can be found in Chock and Jazcilevich & Fuentes-Gea. It is important to note that the results obtained from Eulerian models of this nature are predictions of the ensemble average of the species concentration and not instantaneous values. For this reason, such models are employed in predicting time averaged concentrations.

The main problem in finding a solution to equation $\frac{\partial\langle c_i\rangle}{\partial t}=-\overline{U}\cdot\nabla\langle c_i\rangle-\nabla\cdot\langle c_i'U'\rangle+D\nabla^2\langle c_i\rangle+\langle S_i\rangle$ stems from the radically different character of each of its constituent parts. The advection component is hyperbolic, the turbulent diffusion component, parabolic, and the source/sink term is typically representative of a set of stiff, differential equations. An attempt to obtain a comprehensive solution over each time step is computationally impossible given the magnitude of the problem and current computational limitations. However, general solutions have been proposed by many workers with the aid of simplifying assumptions. Holnicki et al. proposed a solution to the two dimensional atmospheric transport equation using first order closure methods and the finite element method. An account of deposition and chemical kinetics were made though in the form of simple rate equations.

The solution of the one-dimensional atmospheric transport equation using first order closure techniques in conjunction with a combination of finite difference and random choice methods. An economic solution was presented, though without any account of chemistry or deposition phenomena.

Eulerian models suffer the disadvantage that their resolution is confined by the spatial and temporal discretisation of the mesh on which they are solved. The use of a mesh is computationally expensive and traditionally requires some form of optimisation to achieve any degree of efficiency The approach is information rich, providing a description of the relevant transport dynamics at all defined points throughout the domain

Lagrangian Models

The Lagrangian approach to atmospheric dispersion modelling differs from its Eulerian counterparts in that its reference system follows the prevailing vector of atmospheric motion. The term 'Lagrangian' is applied to a wide range of models which simulate pollutant dispersal relative to a shifting reference frame. The general equation of motion, Zanetti, describing the atmospheric dispersion of a single pollutant species, is:

$$\langle c(r,t)\rangle=\int_{-\infty}^{t}\int p(r,t\,|\,r',t')S(r',t')\,dr'dt'$$

where $\langle c(r,t)\rangle$ is the ensemble average concentration at r, at time t, $S(r',t')$ is the source term and $p(r,t\,|\,r',t)$ is the probability density function that an air parcel is moving from r' at t' to r at time t.

Here, the integration in space is carried out across the entire modelling domain while it is assumed that for any r' where t > t' the following expression holds:

$$\int p(r,t \mid r',t')\,dr \le \alpha$$

For those cases where the conservation of mass is enforced, then is assumed to be unity. If phenomena such as chemical reactions and deposition are considered, then the condition 0 < < 1 applies. In applying equation $\langle c(r,t)\rangle = \int_{-\infty}^{t}\int p(r,t \mid r',t')S(r',t')\,dr'dt'$, it is necessary to obtain a suitable description of the probability density function p, where p is solely a function of the prevailing meteorology. Obviously, this assumption applies only to those scenarios where the source emission is limited to the gas phase. If the source emission comprises particulate matter, e.g. dust, ash etc., then particle characteristics have also to be accounted for in p.

A fundamental problem that workers have encountered with equation $\langle c(r,t)\rangle = \int_{-\infty}^{t}\int p(r,t \mid r',t')S(r',t')\,dr'dt'$ and in a wider sense, Lagrangian dispersion models generally, is the relative interpretation of their results. It is fair to assume that the great majority of real-time meteorological and air quality measurements are obtained relative to a stationary reference frame. As a consequence, the results from Lagrangian based models cannot be easily compared with observed measurements. This often presents difficultie during the initial validation and verification phases of model development, as well throughout the post-development period, where simulation results have to be mapped to real life scenarios if they are to be considered 'useful'. This has resulted in a trend where those atmospheric dispersion models which have adopted a Lagrangian approach have typically utilised Lagrangian methods which can be more readily compared to a Eulerian reference frame (e.g. particle/ puff models). Those that have been successfully developed have been required to facilitate some form of Eulerian mapping of their results, in order to achieve a wider application.

Lagrangian Box Models

The most elementary of Lagrangian models is the Lagrangian box model, and is related to the Eulerian box model. The obvious distinction between the two types of box models is that the Lagrangian approach assumes that the rectangular domain is advected horizontally at a rate equal to the local time varying average of the prevailing wind velocity.

The model suffers from the same flaws that are intrinsic to the Eulerian box model, namely a dependency on a constant wind velocity input vector and homogenous and instantaneous mixing. This results in an inability to represent shear layer behaviour and atmospheric flow regimes within the PBL. The Lagrangian box approach increases the degree of spatial information in relation to the pollutant concentrations it predicts though this occurs at the expense of temporal information.

As a consequence of the inherent defects associated with box modelling, the Lagrangian box model has found little application within its field, save for one specific area known as photochemical modelling. Several models have been developed by workers for simulating photochemical reactions within a moving air mass.

Lagrangian Particle Models

Lagrangian particle models embody a breed of models which are essentially part Eulerian, and part Lagrangian in nature. The conceptual view imagines a specified number of inert particles being transported through a defined domain in response to the application of a quantifiable force. Each particle may experience a known variable force and thereby enjoy a unique vector of motion. If the environmental phenomena responsible for particle motion are described using a Eulerian reference frame, then each Lagrangian particle can be referenced to a common origin and ultimately to all particles within the domain. In atmospheric pollutant dispersion models, the environmental forces responsible for instigating motion are attributed to wind flow conditions and gravitational effects. The resultant force of the prevailing wind and gravity provides each particle with an associated velocity component. The trajectory of each particle is expressed as the collective sum of its velocity components over the total trajectory time. With the simulation of hundreds and thousands of particle trajectories, complex advection and diffusion processes can be inferred. This versatility has resulted in an enhanced capability for problem representation, where the attributes time scale associated with individual particles can be varied as required.

Three principal forms of particle model have been suggested by Zanetti and Hockney & Eastwood:

- Particle–Mesh models. The motion of the particle is attributed to environmental forces which are resolved locally using a computational mesh.
- Particle–Particle models. Alternatively, the motion of the particle is calculated solely from the inter-particle forces between all particles within the domain.
- Hybrid models. This approach accounts for combinations of the above where the macro-scale component of motion is provided from a mesh and the micro-scale contribution of motion is resolved from inter-particle reactions.

The particles used by any of these principal types may conceptually represent phenomena ranging from these at molecular levels to 'super-particles'; super-particles being representative of a number of physical particles as is the case in cloud formations. This increases the possible scenarios that a single model may be capable of representing, provided that the rudimentary physics of the problems are compatible.

Particle models employed in the simulation of atmospheric dispersion function in one of two possible modes. The first of these modes is commonly referred to as 'single-particle mode' and considers the motion of each particle as being independent of its co-particles, except when instances of inter-particle collisions occur. A large number of particles, sufficien to generate a representative ensemble average, is required if the dispersion phenomena are to be accurately reproduced. The frame of reference is intrinsically global and interpretation of the results are well suited to a Eulerian frame of reference. In contrast, the 'two-particle mode' describes the diffusion of a number of particles relative to a common, moving, central point within a diffusin cloud of particles. This approach has been employed in Gaussian puff models, where the dispersion behaviour of a single puff is investigated in relation to its centre. If the two-particle mode is to be efficientl related to a Eulerian frame of reference then the trajectory of the particle representing the puff 's centre is required to be independent and to possess information consistent with the single-particle mode.

In either case, a particle will disperse away from its origin of release according to its associated velocity derived from the prevailing atmospheric conditions. The trajectory of each particle is calculated over a predefined time step such that after a time interval D_t has elapsed, the particle moves from its pre-interval position $x(t_1)$ to a new position at $x(t_2)$ given by:

$$x(t_2)=x(t_1)+\int_{t_1}^{t_2} u(x(t),t)dt$$

where u is the local instantaneous wind velocity vector at x(t) and t the trajectory time.

The solution of equation $x(t_2)=x(t_1)+\int_{t_1}^{t_2} u(x(t),t)dt$ requires the description of the local instantaneous wind velocity vector u. This is considered impossible in reality as the wind velocity component exhibits a semi-random behavior resulting from the various contributions of differing scales of atmospheric motions and turbulent eddies. u can be expressed as a resolvable component, and a semi-random component, i.e. $u = \bar{u} + u'$.

The resolvable component $\bar{u}$ is representative of the best estimate of the ensemble average of the Eulerian wind velocity vector. This term is usually calculated from meteorological observations. The diffusivity velocity term u' acts to encompass all those phenomena which are semi-random or subject to variability. It is the description of this term which determines the focus and ultimately the characteristic bias of a particle model. In this respect, u' may be described deterministically using a combination of numerical and empirical methods, or statistically by representing the inherent randomness through statistical inferences and probabilities.

Statistical Approaches to Particle Modeling

In coming to a statistical resolution of u', a stochastic model must be capable of reproducing the randomness of the wind velocity fluctuations. This is typically achieved through some form of weighted random function. A statistical approach to randomness is intuitively natural given that the full spectrum of velocity fluctuations can be described using a single weighted distribution function. The description of the weightings conventionally relies upon Eulerian measurements of u in implying statistical information about u'.

Hanna, Pielke et al. and McNider et al. suggested that both the Eulerian and Lagrangian wind velocity fluctuations could be simulated using a simple Markov process where

$$u'(t_2)=R(\Delta t)u'(t_1)+u''(t_1)$$

where R(Δt) is a vector containing the autocorrelations with lag Δt of the u' component and u" is a purely random vector.

It can be seen that equation $u'(t_2)=R(\Delta t)u'(t_1)+u''(t_1)$ is a function of the previous value of u' and a randomly generated vector. Consequently, velocity vectors with identical initial conditions will typically possess different vector attributes after a single time step. The motion of the particles is independent, resulting in a higher computational efficienc during simulation. Two factors need to be considered before equation $u'(t_2)=R(\Delta t)u'(t_1)+u''(t_1)$ can be successfully applied. The initial condition at $u'(t_0)$ for each particle is required. Historically, the initial conditions are assumed to be a zero vector or random variance derived from Eulerian data. The second factor concerns the suitable evaluation of the Lagrangian autocorrelation function R(Δt), which can be related to the Lagrangian turbulent time scale by:

$$R(\Delta t) = \exp\left[\frac{-\Delta t}{T_L}\right]$$

T_L is a vector containing the Lagrangian time scales for the appropriate number of dimensions and is typically estimated from meteorological measurements. Taking the variances of equation $u'(t_2) = R(\Delta t)u'(t_1) + u''(t_1)$ and rearranging yields:

$$\sigma_{u''} = \sigma_{u'}\left[1 - R^2(\Delta t)\right]^{1/2}$$

If u'' is a purely random vector with independent, normally distributed, zero mean components, then u' is fully described by the distribution su2. The solution of $\sigma_{u''}$ in equation $\sigma_{u''} = \sigma_{u'}\left[1 - R^2(\Delta t)\right]^{1/2}$ rests on finding a suitable description of the distribution $\sigma_{u'}$. This can be achieved using Eulerian wind velocity measurements to obtain an approximate distribution function. The common approach to solving equation $u'(t_2) = R(\Delta t)u'(t_1) + u''(t_1)$ is to employ a Monte Carlo random number generator program in conjunction with equation $\sigma_{u''} = \sigma_{u'}\left[1 - R^2(\Delta t)\right]^{1/2}$ to predict each particle's u'' term. Knowing the autocorrelation function and substituting in the values for u'', the new particle trajectory can be resolved.

This approach is valid only when turbulent conditions within the wind field are stationary, homogeneous and isotropic. Results presented in Ley and Pielke et al. demonstrate that stochastic particle models are capable of accurately representing the dispersion behaviour of airborne pollutants to a very high degree. However, such models often require detailed meteorological information which cannot always be provided. This information typically includes the mean wind velocity and standard deviations of the wind fluctuations in all orthogonal directions

As is the case with the majority of random walk statistical models, an increased number of simulations is required if a representative study is to be established. This necessitates additional analysis of the simulation data and extends the overall simulation period. Further work in this area can be found in Zanetti, Ley and Zanetti.

Deterministic Approaches to Particle Modeling

The deterministic approach to particle modelling assumes the stance that the diffusivity velocity can be derived from considerations of the local species concentration field. Lange suggested a method of calculating the diffusivity velocity by manipulating the differential form of the conservation of mass equation. The method uses velocity values obtained from a Eulerian mesh architecture in conjunction with a linear interpolation scheme to generate localised velocities across the domain. Consider the conservation equation of i species with m mass distributions using a firstorder closure technique

$$\frac{\partial \bar{c}_{im}}{\partial t} = -\left(\bar{U} - v_{g_{im}} k\right)\nabla \bar{c}_{im} - \nabla \cdot K_D \nabla \bar{c}_{im} + S_{im}$$

where,

c = concentration

$i = 1, 2, \ldots, n$ is the number of species

$m = 1, 2, \ldots, n$ the mass distributions

S is the source/sink term

$v_{g_{im}}$ k is the settling velocity due to gravity

K_D is the eddy diffusivity exchange coefficient i.e. the diffusi n of a species concentration in the presence of turbulent eddies.

Equation $\frac{\partial \bar{c}_{im}}{\partial t} = -\left(\bar{U} - v_{g_{im}} k\right)\nabla \bar{c}_{im} - \nabla \cdot K_D \nabla \bar{c}_{im} + S_{im}$ can be rewritten as:

$$\frac{\partial \bar{c}_{im}}{\partial t} + \nabla \cdot \left(\bar{c}_{im}\, U_p\right) = S_{im}$$

given that UP is the particle pseudo velocity $U_P = U_A + U_D$, where U_A is the advection velocity U_A = ui + vj + (w – v_g) k and is obtained directly from a coupled wind field. U_D is the diffusion velocity calculated from the expression:

$$U_D = -\frac{K_D \nabla \bar{c}_{im}}{\bar{c}_{im}}$$

This equation describes the motion of a unit mass particle of concentration c, through the domain by a pseudo velocity U_p. The concentration of the particle is independent of transport and depends only on the effect of the sink/source term with time

The concentration c, specified for each cell at the beginning of the computational cycle, can be used through equation $U_D = -\frac{K_D \nabla \bar{c}_{im}}{\bar{c}_{im}}$, to calculate the diffusive velocity. The outcome is then combined with the advective velocity obtained from the supporting wind flow model, to yield a pseudo velocity for each cell corner within the mesh. Each marker particle presented within a given cell is then transported with a velocity calculated from the linear interpolation of the corner pseudo velocities. The new trajectory of the particle is obtained by forward differencing from the old coordinate. Once all particles have been moved, the new cell concentration can be calculated and the cycle repeats.

The modelling of larger-scale transport events requires the release of a multitude of particles. Each particle is tracked and its position in relation to its Eulerian origin is recorded. Although the wind flow field is derived from a Eulerian grid model, the flow of the particles through the discretised mesh is grid independent. This facilitates a depth of resolution that a mesh architecture is incapable of providing. Converting the grid independent concentration field into a resolution that is compatible with the wind field is far from problematic in any case. The total number of particles (n), of concentration (c), within a given cell, is then averaged to give the total cell concentration (C).

This hybrid particle in cell model was embodied by Lange in the Atmospheric Diffusion Particle In Cell code (ADPIC), utilizing a Eulerian, Mass Adjusted Three Dimensional Wind field model. Preliminary verification demonstrated that model simulations of known analytical solutions generated results that were within 5% of the true solution. For individual case studies, and without site-specific tuning, the model tends to be within a factor of 2 of the observed conditions 60% of the time, while 96% of the time it agreed within an order of magnitude. Unfortunately, the computational technology of the late 1970s prevented Lange from implementing a comprehensive model study in which the sensitivity and performance of ADPIC could be more closely examined.

Work by Sclarew et al. and Amsden predate the work of Lange and deal with particlein- cell modelling generally. Both workers agreed that the Lagrangian particle approach is of a superior nature, surpassing traditional finite difference method in speed, performance and overall efficie

More recently, Lee presented a numerical model for the calculation of three-dimensional flow and pollutant dispersion around structures. The model comprised a finite-element flow code (FEMT-KE) for resolving the wind flow field, while the (ADPIC) code, developed by Lange was employed to predict pollutant dispersion. Again, the model results demonstrated a close correlation between the flow and pollutant dispersion patterns observed near buildings in reality. However, the model has not been consolidated and is complex in its implementation. In addition, Lee makes no account of chemical kinetics and deposition phenomena within the model.

4

Mitigation of Air Pollution

Various air pollution mitigation techniques and devices such as thermal oxidizer, catalytic oxidizer, emissions trading, pollution tax, and dust collection system are discussed in this chapter. This chapter closely examines the globally accepted policies and laws for air pollution to provide an extensive understanding of the subject.

Thermal Oxidizer

Thermal Oxidizers are a class of pollution control devices that use the combustion process to destroy hydrocarbons, volatile organic compounds (VOC), Hydrogen Sulphide, and other compounds as well as smoke particles and odour, hence they are sometimes called afterburners, fume incinerators or tail gas incinerators.

The design of an incineration system is dependent on the pollutant concentration in the waste gas stream, type of pollutant, presence of other gases, level of oxygen, stability of processes vented to the system, and degree of control required. Important design factors include temperature (a temperature high enough to ignite the organic constituents in the waste gas stream), residence time (sufficient time for the combustion reaction to occur), and turbulence or mixing of combustion air with the waste gas. Time, temperature, degree of mixing, and sufficient oxygen concentration govern the completeness of the combustion reaction. Of these, only temperature and oxygen concentration can be significantly controlled after construction. Residence time and mixing are fixed by oxidizer design, and flow rate can be controlled only over a limited range.

The rate at which VOC compounds, volatile HAP, and CO are oxidized is greatly affected by temperature; the higher the temperature, the faster the oxidation reaction proceeds. Because inlet gas concentrations are well below the lower explosive limit (LEL) to prevent preignition explosions in ducting the stream from the process to the oxidizer, the gas must be heated with auxiliary fuel above the autoignition temperature. Thermal destruction of most organics occurs at combustion temperatures between 800°F and 2000°F. Residence time is equal to the oxidizer chamber volume divided by the total actual flow rate of flue gases (waste gas flow, added air, and products of combustion). A residence time of 0.2 to 2.0 seconds, a length-todiameter ratio of 2 to 3 for the chamber dimensions, and an average gas velocity of 10 to 50 feet per second are common. Thorough mixing is necessary to ensure that all waste and fuel come in contact with oxygen.

Because complete mixing generally is not achieved, excess air/oxygen is added (above stoichiometric or theoretical amount) to ensure complete combustion.

Normal operation of a thermal oxidizer should include a fixed outlet temperature or an outlet temperature above a minimum level. A variety of operating parameters that may be used to indicate good operation include: inlet and outlet VOC concentration, outlet combustion temperature, auxiliary fuel input, fuel pressure (magnehelic gauge), fan current (ammeter), outlet CO concentration, and outlet O_2 concentration.

Technologies

Direct Fired Thermal Oxidizer - Afterburner

The simplest technology of thermal oxidation is direct-fired thermal oxidizer. A process stream with hazardous gases is introduced into a firing box through or near the burner and enough residence time is provided to get the desired destruction removal efficiency (DRE) of the VOCs. Most direct-fired thermal oxidizers operate at temperature levels between 980 °C (1,800 °F) and 1,200 °C (2,190 °F) with air flow rates of 0.24 to 24 standard cubic meters per second.

Direct-fired thermal oxidizer using landfill gas as fuel.

Also called afterburners in the cases where the input gases come from a process where combustion is incomplete, these systems are the least capital intensive, and can be integrated with downstream boilers and heat exchangers to optimize fuel efficiency. Thermal Oxidziers are best applied where there is a very high concentration of VOCs to act as the fuel source (instead of natural gas or oil) for complete combustion at the targeted operating temperature.

Regenerative Thermal Oxidizer (RTO)

One of today's most widely accepted air pollution control technologies across industry is a regenerative thermal oxidizer, commonly referred to as a RTO. RTOs use a ceramic bed which is heated from a previous oxidation cycle to preheat the input gases to partially oxidize them. The preheated gases enter a combustion chamber that is heated by an external fuel source to reach the target oxidation temperature which is in the range between 760 °C (1,400 °F) and 820 °C (1,510 °F). The final temperature may be as high as 1,100 °C (2,010 °F) for applications that require maximum destruction. The air flow rates are 2.4 to 240 standard cubic meters per second.

Regenerative thermal oxidizer (RTO) that is 17000 standard cubic feet per minute, or SCFM for short.

Control center with a programmable logic controller for a RTO.

RTOs are very versatile and extremely efficient – thermal efficiency can reach 95%. They are regularly used for abating solvent fumes, odours, etc. from a wide range of industries. Regenerative Thermal Oxidizers are ideal in a range of low to high VOC concentrations up to 10 g/m^3 solvent. There are currently many types Regenerative Thermal Oxidizer on the market with the capabitlity of 99.5+% Volatile Organic Compound (VOC) oxidisation or destruction efficiency. The ceramic heat exchanger(s) in the towers can be designed for thermal efficiencies as high as 97+%.

Ventilation Air Methane Thermal Oxidizer (VAMTOX)

Ventilation air methane thermal oxidizers are used to destroy methane in the exhaust air of underground coal mine shafts. Methane is a greenhouse gas and, when oxidized via thermal combustion, is chemically altered to form CO_2 and H_2O. CO_2 is 25 times less potent than methane when emitted into the atmosphere with regards to global warming. Concentrations of methane in mine ventilation exhaust air of coal and trona mines are very dilute; typically below 1% and often below 0.5%. VAMTOX units have a system of valves and dampers that direct the air flow across one or more ceramic filled bed(s). On start-up, the system preheats by raising the temperature of the heat exchanging ceramic material in the bed(s) at or above the auto-oxidation temperature of methane 1,000 °C (1,830 °F), at which time the preheating system is turned off and mine exhaust air is introduced. Then the methane-filled air reaches the preheated bed(s), releasing the heat from combustion. This heat is then transferred back to the bed(s), thereby maintaining the temperature at or above what is necessary to support auto-thermal operation.

Thermal Recuperative Oxidizer

A less commonly used thermal oxidizer technology is a thermal recuperative oxidizer. Thermal recuperative oxidizers have a primary and/or secondary heat exchanger within the system. A primary heat exchanger preheats the incoming dirty air by recuperating heat from the exiting clean air. This is done by a shell and tube heat exchanger or a plate heat exchanger. As the incoming air passes on one side of the metal tube or plate, hot clean air from the combustion chamber passes on the other side of the tube or plate and heat is transferred to the incoming air through the process of conduction using the metal as the medium of heat transfer. In a secondary heat exchanger the same concept applies for heat transfer, but the air being heated by the outgoing clean process stream is being returned to another part of the plant – perhaps back to the process.

Biomass Fired Thermal Oxidizer

Biomass, such as wood chips, can be used as the fuel for a thermal oxidizer. The biomass is then gasified and the stream with hazardous gases is mixed with the biomassgas in a firing box. Sufficient turbulence, retention time, oxygen content and temperature will ensure destruction of the VOC's. Such biomass fired thermal oxidizer has been installed at Warwick Mills, New Hampshire. The inlet concentrations are between 3000-10.000 ppm VOC. The outlet concentration of VOC are below 3 ppm, thus having a VOC destruction efficiency of 99.8%-99.9%.

Flameless Thermal Oxidizer (FTO)

In a flameless thermal oxidizer system waste gas, ambient air, and auxiliary fuel are premixed prior to passing the combined gaseous mixture through a preheated inert ceramic media bed. Through the transfer of heat from the ceramic media to the gaseous mixture the organic compounds in the gas are oxidized to innocuous byproducts, i.e., carbon dioxide (CO_2) and water vapor (H_2O) while also releasing heat into the ceramic media bed.

The gas mixture temperature is kept below the lower flammability limit based on the percentages of each organic species present. Flameless thermal oxidizers are designed to operate safely and reliably below the composite LFL while maintaining a constant operating temperature. Waste gas streams experience multiple seconds of residence time at high temperatures leading to measured destruction removal efficiencies that exceed 99.9999%. Premixing all of the gases prior to treatment eliminates localized high temperatures which leads to thermal NOx typically below 2 ppmV. Flameless thermal oxidizer technology was originally developed at the U.S. Department of Energy to more efficiently convert energy in burners, process heaters, and other thermal systems.

Catalytic Oxidizer

Catalytic oxidizer (also known as catalytic incinerator) is another category of oxidation systems that is similar to typical thermal oxidizers, but the catalytic oxidizers use a catalyst to promote the oxidation. Catalytic oxidation occurs through a chemical reaction between the VOC hydrocarbon molecules and a precious-metal catalyst bed that is internal to the oxidizer system. A catalyst is a substance that is used to accelerate the rate of a chemical reaction, allowing the reaction to occur in a normal temperature range between 340 °C (644 °F) and 540 °C (1,004 °F).

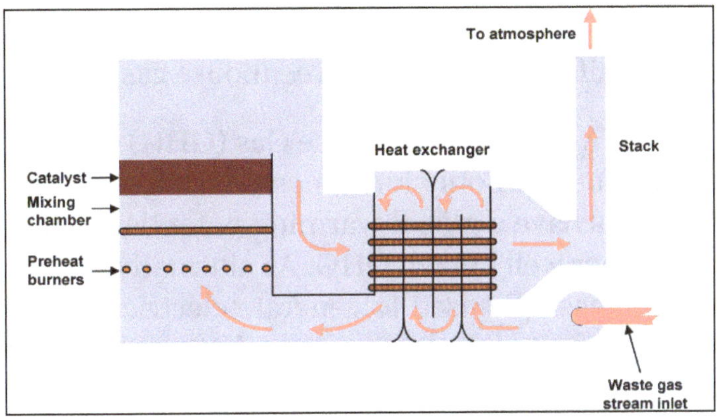

Schematic of Recuperative Catalytic Oxidizer.

Regenerative Catalytic Oxidizer (RCO)

The catalyst can be used in a Regenerative Thermal Oxidizer (RTO) to allow lower operating temperatures. This is also called Regenerative Catalytic Oxidizer or RCO. For example, the thermal ignition temperature of carbon monoxide is normally 609 °C (1,128 °F). By utilizing a suitable oxidation catalyst, the ignition temperature can be reduced to around 200 °C (392 °F). This can result in lower operating costs than a RTO. Most systems operate within the 260 °C (500 °F) to 1,000 °C (1,830 °F) degree range. Some systems are designed to operate both as RCOs and RTOs. When these systems are used special design considerations are utilized to reduce the probability of overheating (dilution of inlet gas or recycling), as these high temperatures would deactivate the catalyst, e.g. by sintering of the active material.

Recuperative Catalytic Oxidizer

Catalytic oxidizers can also be in the form of recuperative heat recovery to reduce the fuel requirement. In this form of heat recovery, the hot exhaust gases from the oxidizer pass through an heat exchanger to heat the new incoming air to the oxidizer.

Ventilation Air Methane Thermal Oxidizer

Ventilation air methane thermal oxidizers (or VAMTOX) are a type of processing equipment used for greenhouse gas abatement related to underground mining operations that destroys gaseous methane at a high temperature.

Principle

Ventilation Air Methane Thermal Oxidizers are used to destroy methane in the exhaust air of underground coal mine shafts. Methane is a greenhouse gas that burns to form carbon dioxide (CO_2) and water vapour. Carbon dioxide is 25 times less potent than methane when emitted into the atmosphere with regards to global warming. Concentrations of methane in ventilation exhaust air of coal and trona mines are very dilute; typically below 1% and often below 0.5%. Flow rates are so high that ventilation air methane constitutes the largest source of methane emissions at most mines. This methane emission wastes energy and contributes significantly to global greenhouse gas (GHG) emissions.

Ventilation Air Methane (VAM) is a Greenhouse Gas (GHG) released by coal or mineral mining operations, it is also commonly known as Coal Mine Methane (CMM). The air pollutants from this process have a global warming potential 21 times greater than carbon dioxide, which is another well known GHG. As alternative energy technologies have developed, methane is increasingly used to generate electricity and heat. Landfills and mines represent two of the most abundant sources, however reports indicate that more than 50% of all VAM is exhausted from mine ventilation systems and remains unutilized.

Mining operations can profit by not venting the methane to atmosphere if they incorporate the proper VAM oxidation technology for air purification and combine it with a heat-to-power system for electricity production. These projects can often be funded without using extensive capital resources when the carbon credit and electrical power potential is taken into account.

Based on the general process conditions, a Regenerative Thermal Oxidizer (RTO) is clearly the best abatement device for VAM but it is certainly not a standard RTO application. These systems must be built conservatively and with great flexibility to ensure continuous operation. Most mine operators demand that the emission control system and related hooding for collection have zero impact on the operation and employee safety.

The sheer volume of air coming from the mine shafts is massive and can vary greatly. Therefore VAM oxidizers must handle the large volume / low concentration stream, yet be capable of adapting to emission and flow spikes.

Regenerative Thermal Oxidizer

Air Pollution Control using a Regenerative Thermal Oxidizer(RTO) is most appropriate in process applications with lower concentrations (up to 10% LEL) of Volatile Organic

Compounds (VOC's) and air flows higher than 5000 SCFM. The chief benefit of installing a Regenerative Thermal Oxidizer Media System is lower operating costs when compared to other emission control technologies. High energy efficiency, around 95%, is achieved by recovering and re-using the excess heat energy which is created in the process of combusting the organic materials contained in the air stream. It is possible to obtain destruction efficiency in excess of 99% in some cases.

Pollutant Destruction Phase

1. The process begins with a high-pressure supply fan that forces exhaust fumes into the Regenerative Thermal Oxidizer.

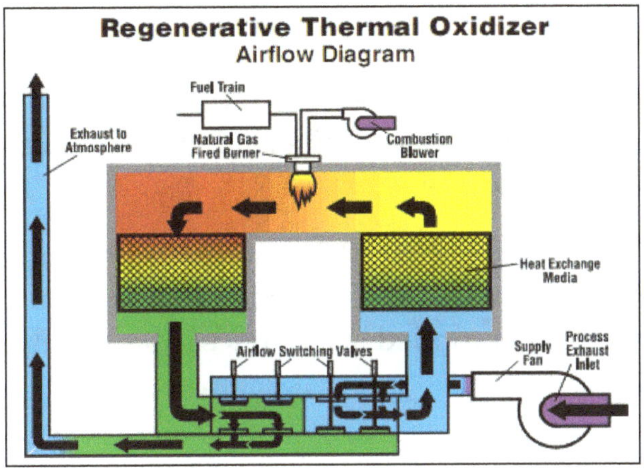

2. Inlet switch valves direct the pollutant-laden air into one of the energy recovery canisters.
3. The air continues to travel from the valve assembly up through the first heat exchanger canister where heat is absorbed from the ceramic media.
4. The preheated air moves into the combustion chamber and is held at an elevated temperature close to that required for oxidation.
5. The pollutant is destroyed while in the combustion chamber.

The Clean Air Cycle

6. Once out of the combustion chamber, the clean, hot air moves through a second energy recovery canister where the ceramic media absorbs the heat generated during thermal oxidation.
7. Finally, the cooled, clean air is routed into the atmosphere through outlet switch values and ultimately the exhaust stack.
   - *Thermal energy efficiencies range from 85-97%.*
   - *Destruction efficiencies of 99% are typically guaranteed.*

Best Available Technology

'Best available techniques' (BAT) means the available techniques which are the best for preventing or minimising emissions and impacts on the environment. You need to use BAT if your operation is an installation (eg a facility that carries out an industrial process like a refinery, food factory or intensive farm).

'Techniques' include both the technology used and the way your installation is designed, built, maintained, operated and decommissioned.

The European Commission produces best available technique reference documents or BREF notes. They contain 'best available techniques' (BAT) for installations.

For example, there's a BREF for intensive agriculture which contains BAT for housing for pig rearing units and a BREF for the textiles industry which contains BAT for selecting materials for textile manufacture.

The European Commission is updating BREF notes and the updated versions also include 'BAT conclusion documents'. These contain emission limits associated with BAT ('BAT AELs') which must be complied with unless the Environment Agency agrees you've met certain criteria.

Ways to follow BAT

Your permit conditions may tell you what BAT you must use or they may set emission limit values (ELV) or other environmental outcomes, based on BAT.

If your permit says you must follow BAT or 'appropriate measures' to achieve an outcome or ELV, check the guide for your activity for the BAT for that process.

You may have to decide which BAT to use yourself if your permit doesn't tell you which BAT to use.

You may also need to take additional measures to meet the conditions in your permit.

BAT in your Permit Application

When you apply for an environmental permit you must state whether you're going to follow each BAT that applies to your activity, or propose an alternative.

You need to do this in the 'operating techniques' section of the application form.

For BAT that you're proposing to follow, you must explain how you're going to either:

- follow the BAT conclusions and meet the BAT-associated emissions level (for BAT that are contained in BAT conclusions).

- follow the BREF note and the technical guidance for activities that don't have BAT conclusions.

The origin of the "best available technology" is derived from the second half of the 1800s. It has been expressed in various ways, such as "Best practicable means", but has always had basically the same meaning. Today's "Best Available Technology" was coined for the first time in the 1992 OSPAR Convention where and accurately described by:

1. The use of the best available techniques shall emphasise the use of non-waste technology, if available.
2. The term "best available techniques" means the latest stage of development (state of the art) of processes, of facilities or of methods of operation which indicate the practical suitability of a particular measure for limiting discharges, emissions and waste. In determining whether a set of processes, facilities and methods of operation constitute the best available techniques in general or individual cases, special consideration shall be given to:
   (a) comparable processes, facilities or methods of operation which have recently been successfully tried out;
   (b) technological advances and changes in scientific knowledge and understanding;
   (c) the economic feasibility of such techniques;
   (d) time limits for installation in both new and existing plants;
   (e) the nature and volume of the discharges and emissions concerned.
3. It therefore follows that what is "best available techniques" for a particular process will change with time in the light of technological advances, economic and social factors, as well as changes in scientific knowledge and understanding.
4. If the reduction of discharges and emissions resulting from the use of best available techniques does not lead to environmentally acceptable results, additional measures have to be applied.

European Union Directives

Best available techniques not entailing excessive costs (BATNEEC), sometimes referred to as *best available technology*, was introduced in 1984 with Directive 84/360/EEC and applied to air pollution emissions from large industrial installations.

In 1996, Directive 84/360/EEC was superseded by the Integrated pollution prevention and control directive (IPPC), 96/61/EC, which applied the framework concept of *Best Available Techniques* (BAT) to the integrated control of pollution to the three media air, water and soil. The concept is also part of the directive's recast in 2008 (2008/1/EC) and its successor directive, the Industrial Emissions Directive 2010/75/EU published in 2010.

According to article 15(2) of the Industrial Emissions Directive, emission limit values and the equivalent parameters and technical measures in permits shall be based on the best available techniques, without prescribing the use of any technique or specific technology.

The directive includes a definition of best available techniques in article 3(10):

"best available techniques" means the most effective and advanced stage in the development of activities and their methods of operation which indicates the practical suitability of particular techniques for providing the basis for emission limit values and other permit conditions designed to prevent and, where that is not practicable, to reduce emissions and the impact on the environment as a whole:

- "Techniques" includes both the technology used and the way in which the installation is designed, built, maintained, operated and decommissioned.
- "Available" means those developed on a scale which allows implementation in the relevant industrial sector, under economically and technically viable conditions, taking into consideration the costs and advantages, whether or not the techniques are used or produced inside the Member State in question, as long as they are reasonably accessible to the operator.
- "Best" means most effective in achieving a high general level of protection of the environment as a whole.

  BAT for a given industrial sector are described in BAT reference documents (BREFs) as defined in article 3(11) of the Industrial Emissions Directive. BREFs are the result of an exchange of information between European Union Member States, the industries concerned, non-governmental organisations promoting environmental protection and the European Commission pursuant to article 13 of the directive. This exchange of information is often called the Sevilla process because it is steered by the Institute for Prospective Technological Studies of the European Commissions' Joint Research Centre, which is based in Seville. The process is described in detail in Commission Implementing Decision 2012/119/EU.The most important chapter of the BREFs, the BAT conclusions, are published as implementing decisions of the European Commission in the Official Journal of the European Union. According to article 14(3) of the Industrial Emissions Directive, the BAT conclusions shall be the reference for setting permit conditions of large industrial installations.

United States Environmental law

Clean Air Act

The Clean Air Act requires that certain facilities employ Best Available Control Technology to control emissions.

> An emission limitation based on the maximum degree of reduction of each pollutant subject to regulation under this Act emitted from or which results from any major emitting facility, which the permitting authority, on a case-by-case basis, taking into account energy, environmental, and economic impacts and other costs, determines is achievable for such facility through application of production processes and available methods, systems, and techniques, including fuel cleaning, clean fuels, or treatment or innovative fuel combustion techniques for control of each such pollutant.

Clean Water Act

The Clean Water Act (CWA) requires issuance of national industrial wastewater discharge regulations (called "effluent guidelines"), which are based on BAT and several related standards.

> Effluent limitations for categories and classes of point sources, which (i) shall require application of the best available technology economically achievable for such category or class, which will result in reasonable further progress toward the national goal of eliminating the discharge of all pollutants. Factors relating to the assessment of best available technology shall take into account the age of equipment and facilities involved, the process employed, the engineering aspects of the application of various types of control techniques, process changes, the cost of achieving such effluent reduction, non-water quality environmental impact (including energy requirements), and such other factors as the Administrator deems appropriate.

> In the development of the effluent standards, the BAT concept is a "model" technology rather than a specific regulatory requirement. The U.S. Environmental Protection Agency(EPA) identifies a particular model technology for an industry, and then writes a regulatory performance standard based on the model. The performance standard is typically expressed as a numeric effluent limit measured at the discharge point. The industrial facility may use any technology that meets the performance standard.

> A related CWA provision for cooling water intake structures requires standards based on "best technology available."

> The location, design, construction, and capacity of cooling water intake structures reflect the best technology available for minimizing adverse environmental impact.

International Conventions

The concept of BAT is also used in a number of international conventions such as the Minamata Convention on Mercury, the Stockholm Convention on Persistent Organic Pollutants, or the OSPAR Convention for the protection of the marine environment of the North-East Atlantic.

Emissions Trading

Emissions trading is a market-based approach to controlling pollution. By creating tradable pollution permits it attempts to add the profit motive as an incentive for good performance, unlike traditional environmental regulation based solely on the threat of penalties.

Power plants in Europe, such as this coal power station in Gelsenkirchen, Germany, are regulated by an emissions trading scheme.

Developed in the 70s and 80s, emissions trading was introduced in the US in 1990 to combat acid rain, but more recently it has grown in prominence as a way of tackling greenhouse gas emissions linked to climate change.

The main form of emissions trading is known as "cap and trade": a cap on emissions is set and then permits are created up to the level of this cap. The companies or other entities covered by the scheme need to hold one permit for every tonne of pollution (CO2e) they emit. Allowing a trade in these permits puts a price on pollution – the cost of emitting one tonne of carbon dioxide is the cost of the permit – and creates flexibility as to how and where pollution is reduced.

The theory is that setting a limit on pollution and allowing the market to decide how to stay within that limit is ideally suited to reducing carbon emissions, which come from almost all forms of economic activity and mix into the atmosphere with global effect. The market should ensure that the emissions cuts happen at the lowest possible cost, and the cap can be lowered year by year in a managed way.

Supporters argue that this is preferable to other forms of pricing, such as carbon taxes, which do not guarantee any particular level of reduction. However, critics often emphasise the degree to which emissions trading has been marred by weak caps, free handouts of permits to the biggest polluters and the purchase of "offsets" – carbon credits bought from outside the cap-and-trade system from carbon reduction projects in the developing world.

Emissions trading is a central element of the Kyoto protocol in the form of the Clean Development Mechanism (CDM) and is the cornerstone policy of the EU, whose Emissions Trading System (ETS) is the largest in the world. The expansion of emissions trading was slowed significantly by the US decision to abandon a proposed national policy, although groups of states have set up regional schemes. A number of countries are considering the adoption of some form of cap and trade, including China and South Korea.

A coal power plant in Germany. Due to emissions trading, coal may become a less competitive fuel than other options.

Pollution is the prime example of a market externality. An externality is an effect of some activity on an entity (such as a person) that is not party to a market transaction related to that activity. Emissions trading is a market-based approach to address pollution. The overall goal of an emissions trading plan is to minimize the cost of meeting a set emissions target.

In an emissions trading system, the government sets an overall limit on emissions, and defines permits (also called allowances), or limited authorizations to emit, up to the level of the overall limit. The government may sell the permits, but in many existing schemes, it gives permits to participants (regulated polluters) equal to each participant's baseline emissions. The baseline is determined by reference to the participant's historical emissions. To demonstrate compliance, a participant must hold permits at least equal to the quantity of pollution it actually emitted during the time period. If every participant complies, the total pollution emitted will be at most equal to the sum of individual limits. Because permits can be bought and sold, a participant can choose either to use its permits exactly (by reducing its own emissions); or to emit less than its permits, and perhaps sell the excess permits; or to emit more than its permits, and buy permits from other participants. In effect, the buyer pays a charge for polluting, while the seller gains a reward for having reduced emissions.

In many schemes, organizations which do not pollute (and therefore have no obligations) may also trade permits and financial derivatives of permits. In some schemes, participants can bank allowances to use in future periods. In some schemes, a proportion of all traded permits must be retired periodically, causing a net reduction in emissions over time. Thus, environmental groups may buy and retire permits, driving up the price of the remaining permits according to the law of demand. In most schemes, permit owners can donate permits to a nonprofit entity and receive a tax deduction. Usually, the government lowers the overall limit over time, with an aim towards a national emissions reduction target.

According to the Environmental Defense Fund, cap-and-trade is the most environmentally and economically sensible approach to controlling greenhouse gas emissions, the primary cause of global warming, because it sets a limit on emissions, and the trading encourages companies to innovate in order to emit less.

"International trade can offer a range of positive and negative incentives to promote international cooperation on climate change (robust evidence, medium agreement). Three issues are key to developing constructive relationships between international trade and climate agreements: how existing trade policies and rules can be modified to be more climate friendly; whether border adjustment measures (BAMs) or other trade measures can be effective in meeting the goals of international climate agreements; whether the UNFCCC, World Trade Organization (WTO), hybrid of the two, or a new institution is the best forum for a trade-and-climate architecture."

Market and Least-cost

Some economists have urged the use of market-based instruments such as emissions trading to address environmental problems instead of prescriptive "command-and-control" regulation. Command and control regulation is criticized for being insensitive to geographical and technological differences, and therefore inefficient.; however, this is not always so, as shown by the WW-II rationing program in the U.S. in which local and regional boards made adjustments for these differences.

After an emissions limit has been set by a government political process, individual companies are free to choose how or whether to reduce their emissions. Failure to report emissions and surrender emission permits is often punishable by a further government regulatory mechanism, such as a fine that increases costs of production. Firms will choose the least-cost way to comply with the pollution regulation, which will lead to reductions where the least expensive solutions exist, while allowing emissions that are more expensive to reduce.

Under an emissions trading system, each regulated polluter has flexibility to use the most cost-effective combination of buying or selling emission permits, reducing its emissions by installing cleaner technology, or reducing its emissions by reducing production. The most cost-effective strategy depends on the polluter's marginal abatement cost and the market price of permits. In theory, a polluter's decisions should lead to an economically efficient allocation of reductions among polluters, and lower compliance costs for individual firms and for the economy overall, compared to command-and-control mechanisms.

Emission Markets

For emissions trading where greenhouse gases are regulated, one emissions permit is considered equivalent to one metric ton of carbon dioxide (CO_2) emissions. Other names for emissions permits are carbon credits, Kyoto units, assigned amount units, and Certified Emission Reduction units (CER). These permits can be sold privately or in the international market at the prevailing market price. These trade and settle internationally, and hence allow permits to be transferred between countries. Each international transfer is validated by the United Nations Framework Convention on Climate Change (UNFCCC). Each transfer of ownership within the European Union is additionally validated by the European Commission.

Emissions trading programmes such as the European Union Emissions Trading System (EU ETS) complement the country-to-country trading stipulated in the Kyoto Protocol by allowing private trading of permits. Under such programmes – which are generally co-ordinated with the national emissions targets provided within the framework of the Kyoto Protocol – a national or international authority allocates permits to individual companies based on established criteria, with a view to meeting national and/or regional Kyoto targets at the lowest overall economic cost.

Trading exchanges have been established to provide a spot market in permits, as well as futures and options market to help discover a market price and maintain liquidity. Carbon prices are normally quoted in euros per tonne of carbon dioxide or its equivalent (CO_2e). Other greenhouse gases can also be traded, but are quoted as standard multiples of carbon dioxide with respect to their global warming potential. These features reduce the quota's financial impact on business, while ensuring that the quotas are met at a national and international level.

Currently, there are six exchanges trading in UNFCCC related carbon credits: the Chicago Climate Exchange (until 2010), European Climate Exchange, NASDAQ OMX Commodities Europe, PowerNext, Commodity Exchange Bratislava and the European Energy Exchange. NASDAQ OMX Commodities Europe listed a contract to trade offsets generated by a CDM carbon project called Certified Emission Reductions. Many companies now engage in emissions abatement, offsetting, and sequestration programs to generate credits that can be sold on one of the exchanges. At least one private electronic market has been established in 2008: CantorCO2e. Carbon credits at Commodity Exchange Bratislava are traded at special platform called Carbon place.

Trading in emission permits is one of the fastest-growing segments in financial services in the City of London with a market estimated to be worth about €30 billion in 2007. Louis Redshaw, head of environmental markets at Barclays Capital, predicts that "carbon will be the world's biggest commodity market, and it could become the world's biggest market overall."

Pollution Markets

An emission license directly confers a right to emit pollutants up to a certain rate. In contrast, a pollution license for a given location confers the right to emit pollutants at a rate which will cause no more than a specified increase at the pollution-level. For concreteness, consider the following model.

- There are n agents each of which emits e_i pollutants.
- There are m locations each of which suffers pollution q_i.
- The pollution is a linear combination of the emissions. The relation between e and q is given by a *diffusion matrix* H, such that: $q = H \cdot e$.

As an example, consider three countries along a river (as in the fair river sharing setting).

- Pollution in the upstream country is determined only by the emission of the upstream country. $q_1 = e_1$.
- Pollution in the middle country is determined by its own emission and by the emission of country 1: $q_2 = e_1 + e_2$.
- Pollution in the downstream country is the sum of all emissions: $q_3 = e_1 + e_2 + e_3$.

So the matrix H in this case is a triangular matrix of ones.

Each pollution-license for location i permits its holder to emit pollutants that will cause at most this level of pollution at location i. Therefore, a polluter that affects water quality at a number of points has to hold a portfolio of licenses covering all relevant monitoring-points. In the above example, if country 2 wants to emit a unit of pollutant, it should purchase two permits: one for location 2 and one for location 3.

Montgomery shows that, while both markets lead to efficient license allocation, the market in pollution-licenses is more widely applicable than the market in emission-licenses.

Comparison with other Methods of Emission Reduction

Cap and trade is the textbook emissions trading program. Other market-based approaches include baseline-and-credit, and pollution tax. They all put a price on pollution, and so provide an economic incentive to reduce pollution beginning with the lowest-cost opportunities. By contrast, in a command-and-control approach, a central authority designates pollution levels each facility is allowed to emit. Cap and trade essentially functions as a tax where the tax rate is variable based on the relative cost of abatement per unit, and the tax base is variable based on the amount of abatement needed.

Baseline and Credit

In a baseline and credit program, polluters can create permits, called credits or offsets, by reducing their emissions below a baseline level, which is often the historical emissions level from a designated past year. Such credits can be bought by polluters that have a regulatory limit.

Pollution Tax

Emissions fees or environmental tax is a surcharge on the pollution created while producing goods and services. For example, a carbon tax is a tax on the carbon content of fossil fuels that aims to discourage their use and thereby reduce carbon dioxide emissions. The two approaches are overlapping sets of policy designs. Both can have a range of scopes, points of regulation, and price schedules. They can be fair or unfair, depending on how the revenue is used. Both have the effect of increasing the price of goods (such as fossil fuels) to consumers. A comprehensive, upstream, auctioned cap-and-trade system is very similar to a comprehensive, upstream carbon tax. Yet, many commentators sharply contrast the two approaches.

The main difference is what is defined and what derived. A tax is a price control, while cap-and-trade method acts is a quantity control instrument. That is, a tax is a unit price for pollution that is set by authorities, and the market determines the quantity emitted; in cap and trade, authorities determine the amount of pollution, and the market determines the price. This difference affects a number of criteria.

Responsiveness to inflation: Cap-and-trade has the advantage that it adjusts to inflation (changes to overall prices) automatically, while emissions fees must be changed by regulators.

Responsiveness to cost changes: It is not clear which approach is better. It is possible to combine the two into a safety valve price: a price set by regulators, at which polluters can buy additional permits beyond the cap.

Responsiveness to recessions: This point is closely related to responsiveness to cost changes, because recessions cause a drop in demand. Under cap and trade, the emissions cost automatically decreases, so a cap-and-trade scheme adds another automatic stabilizer to the economy - in effect, an automatic fiscal stimulus. However, a lower pollution price also results in reduced efforts to reduce pollution. If the government is able to stimulate the economy regardless of the cap-and-trade scheme, an excessively low price causes a missed opportunity to cut emissions faster than planned. Instead, it might be better to have a price floor (a tax). This is especially true when cutting pollution is urgent, as with greenhouse gas emissions. A price floor also provides certainty and stability for investment in emissions reductions: recent experience from the UK shows that nuclear power operators are reluctant to invest on "un-subsidised" terms unless there is a guaranteed price floor for carbon (which the EU emissions trading scheme does not presently provide).

Responsiveness to uncertainty: As with cost changes, in a world of uncertainty, it is not clear whether emissions fees or cap-and-trade systems are more efficient—it depends on how fast the marginal social benefits of reducing pollution fall with the amount of cleanup (e.g., whether inelastic or elastic marginal social benefit schedule).

Other: The magnitude of the tax will depend on how sensitive the supply of emissions is to the price. The permit price of cap-and-trade will depend on the pollutant market. A tax generates government revenue, but full-auctioned emissions permits can do the same. A similar upstream cap-and-trade system could be implemented. An upstream carbon tax might be the simplest to administer. Setting up a complex cap-and-trade arrangement that is comprehensive has high institutional needs.

Command-and-control Regulation

Command and control is a system of regulation that prescribes emission limits and compliance methods for each facility or source. It is the traditional approach to reducing air pollution.

Command-and-control regulations are more rigid than incentive-based approaches such as pollution fees and cap and trade. An example of this is a performance standard which sets an emissions goal for each polluter that is fixed and, therefore, the burden of reducing pollution cannot be shifted to the firms that can achieve it more cheaply. As a result, performance standards are likely to be more costly overall. The additional costs would be passed to end consumers.

Economics of International Emissions Trading

It is possible for a country to reduce emissions using a Command-Control approach, such as regulation, direct and indirect taxes. The cost of that approach differs between countries because the Marginal Abatement Cost Curve (MAC) — the cost of eliminating an additional unit of pollution — differs by country. It might cost China \$2 to eliminate a ton of CO_2, but it would probably cost Norway or the U.S. much more. International emissions-trading markets were created precisely to exploit differing MACs.

Example

Emissions trading through *Gains from Trade* can be more beneficial for both the buyer and the seller than a simple emissions capping scheme.

Consider two European countries, such as Germany and Sweden. Each can either reduce all the required amount of emissions by itself or it can choose to buy or sell in the market.

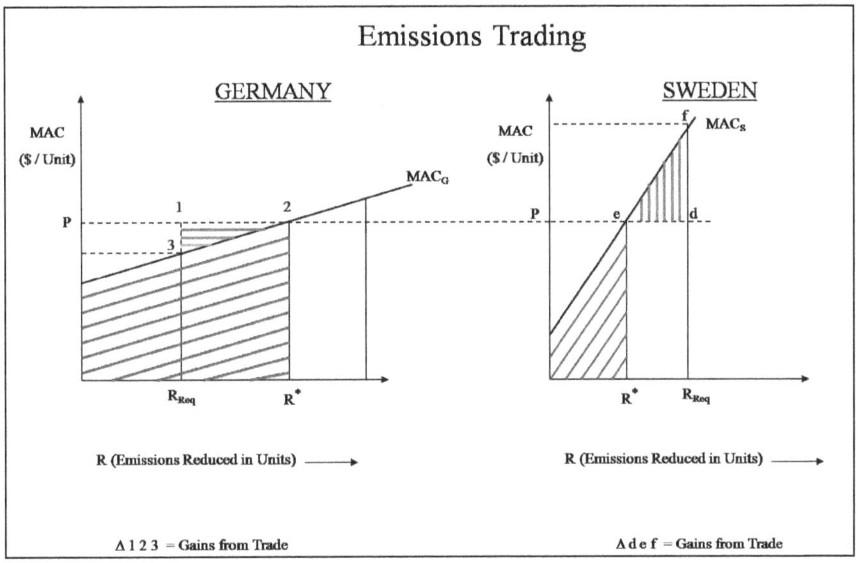

Example MACs for two different countries.

Suppose Germany can abate its CO_2 at a much cheaper cost than Sweden, i.e. $MAC_S > MAC_G$ where the MAC curve of Sweden is steeper (higher slope) than that of Germany, and R_{Req} is the total amount of emissions that need to be reduced by a country.

On the left side of the graph is the MAC curve for Germany. R_{Req} is the amount of required reductions for Germany, but at R_{Req} the MAC_G curve has not intersected the market emissions permit price of CO_2 (market permit price = P = λ). Thus, given the market price of CO_2 allowances, Germany has potential to profit if it abates more emissions than required.

On the right side is the MAC curve for Sweden. R_{Req} is the amount of required reductions for Sweden, but the MAC_S curve already intersects the market price of CO_2 permits before R_{Req} has been reached. Thus, given the market price of CO_2 permits, Sweden has potential to make a cost saving if it abates fewer emissions than required internally, and instead abates them elsewhere.

In this example, Sweden would abate emissions until its MAC_S intersects with P (at R*), but this would only reduce a fraction of Sweden's total required abatement.

After that it could buy emissions credits from Germany for the price *P* (per unit). The internal cost of Sweden's own abatement, combined with the permits it buys in the market from Germany, adds up to the total required reductions (R_{Req}) for Sweden. Thus Sweden can make a saving from buying permits in the market (Δ d-e-f). This represents the "Gains from Trade", the amount of additional expense that Sweden would otherwise have to spend if it abated all of its required emissions by itself without trading.

Germany made a profit on its additional emissions abatement, above what was required: it met the regulations by abating all of the emissions that was required of it (R_{Req}). Additionally, Germany sold its surplus permits to Sweden, and was paid *P* for every unit it abated, while spending less than *P*. Its total revenue is the area of the graph (R_{Req} 1 2 R*), its total abatement cost is area (R_{Req} 3 2 R*), and so its net benefit from selling emission permits is the area (Δ 1-2-3) i.e. Gains from Trade.

The two R* (on both graphs) represent the efficient allocations that arise from trading.

- Germany: sold (R* - R_{Req}) emission permits to Sweden at a unit price *P*.
- Sweden bought emission permits from Germany at a unit price *P*.

If the total cost for reducing a particular amount of emissions in the *Command Control* scenario is called *X*, then to reduce the same amount of combined pollution in Sweden and Germany, the total abatement cost would be less in the *Emissions Trading* scenario i.e. (X — Δ 123 - Δ def).

The example above applies not just at the national level, but also between two companies in different countries, or between two subsidiaries within the same company.

Applying the Economic Theory

The nature of the pollutant plays a very important role when policy-makers decide which framework should be used to control pollution. CO_2 acts globally, thus its impact on the environment is generally similar wherever in the globe it is released. So the location of the originator of the emissions does not matter from an environmental standpoint.

The policy framework should be different for regional pollutants (e.g. SO_2 and NO_x, and also mercury) because the impact of these pollutants may differ by location. The same amount of a regional pollutant can exert a very high impact in some locations and a low impact in other locations, so it matters where the pollutant is released. This is known as the *Hot Spot* problem.

A Lagrange framework is commonly used to determine the least cost of achieving an objective, in this case the total reduction in emissions required in a year. In some cases, it is possible to use the Lagrange optimization framework to determine the required reductions for each country (based on their MAC) so that the total cost of reduction is minimized. In such a scenario, the Lagrange multiplier represents the market allowance price (P) of a pollutant, such as the current market price of emission permits in Europe and the USA.

Countries face the permit market price that exists in the market that day, so they are able to make individual decisions that would minimize their costs while at the same time achieving regulatory compliance. This is also another version of the Equi-Marginal Principle, commonly used in economics to choose the most economically efficient decision.

Prices Versus Quantities and the Safety Valve

There has been longstanding debate on the relative merits of *price* versus *quantity* instruments to achieve emission reductions.

An emission cap and permit trading system is a *quantity* instrument because it fixes the overall emission level (quantity) and allows the price to vary. Uncertainty in future supply and demand conditions (market volatility) coupled with a fixed number of pollution permits creates an uncertainty in the future price of pollution permits, and the industry must accordingly bear the cost of adapting to these volatile market conditions. The burden of a volatile market thus lies with the industry rather than the controlling agency, which is generally more efficient. However, under volatile market conditions, the ability of the controlling agency to alter the caps will translate into an ability to pick "winners and losers" and thus presents an opportunity for corruption.

In contrast, an emission tax is a *price* instrument because it fixes the price while the emission level is allowed to vary according to economic activity. A major drawback of an emission tax is that the environmental outcome (e.g. a limit on the amount of emissions) is not guaranteed. On one hand, a tax will remove capital from the industry, suppressing possibly useful economic activity, but conversely, the polluter will not need to hedge as much against future uncertainty since the amount of tax will track with profits. The burden of a volatile market will be borne by the controlling (taxing) agency rather than the industry itself, which is generally less efficient. An advantage is that, given a uniform tax rate and a volatile market, the taxing entity will not be in a position to pick "winners and losers" and the opportunity for corruption will be less.

Assuming no corruption and assuming that the controlling agency and the industry are equally efficient at adapting to volatile market conditions, the best choice depends on the sensitivity of the costs of emission reduction, compared to the sensitivity of the benefits (i.e., climate damage avoided by a reduction) when the level of emission control is varied.

Because there is high uncertainty in the compliance costs of firms, some argue that the optimum choice is the price mechanism. However, the burden of uncertainty cannot be eliminated, and in this case it is shifted to the taxing agency itself.

The overwhelming majority of climate scientists have repeatedly warned of a threshold in atmospheric concentrations of carbon dioxide beyond which a run-away warming effect could take place, with a large possibility of causing irreversible damage.

With such a risk, a quantity instrument may be a better choice because the quantity of emissions may be capped with more certainty. However, this may not be true if this risk exists but cannot be attached to a known level of greenhouse gas (GHG) concentration or a known emission pathway.

A third option, known as a *safety valve*, is a hybrid of the price and quantity instruments. The system is essentially an emission cap and permit trading system but the maximum (or minimum) permit price is capped. Emitters have the choice of either obtaining permits in the marketplace or buying them from the government at a specified trigger price (which could be adjusted over time). The system is sometimes recommended as a way of overcoming the fundamental disadvantages of both systems by giving governments the flexibility to adjust the system as new information comes to light. It can be shown that by setting the trigger price high enough, or the number of permits low enough, the safety valve can be used to mimic either a pure quantity or pure price mechanism.

All three methods are being used as policy instruments to control greenhouse gas emissions: the EU-ETS is a *quantity* system using the cap and trading system to meet targets set by National Allocation Plans; Denmark has a price system using a carbon tax (World Bank, 2010, p. 218), while China uses the CO_2 market price for funding of its Clean Development Mechanism projects, but imposes a *safety valve* of a minimum price per tonne of CO_2.

Carbon Leakage

Carbon leakage is the effect that regulation of emissions in one country/sector has on the emissions in other countries/sectors that are not subject to the same regulation. There is no consensus over the magnitude of long-term carbon leakage.

In the Kyoto Protocol, Annex I countries are subject to caps on emissions, but non-Annex I countries are not. Barker *et al.* (2007) assessed the literature on leakage. The leakage rate is defined as the increase in CO_2 emissions outside the countries taking domestic mitigation action, divided by the reduction in emissions of countries taking domestic mitigation action. Accordingly, a leakage rate greater than 100% means that actions to reduce emissions within countries had the effect of increasing emissions in other countries to a greater extent, i.e., domestic mitigation action had actually led to an increase in global emissions.

Estimates of leakage rates for action under the Kyoto Protocol ranged from 5% to 20% as a result of a loss in price competitiveness, but these leakage rates were considered very uncertain. For energy-intensive industries, the beneficial effects of Annex I actions through technological development were considered possibly substantial. However, this beneficial effect had not been reliably quantified. On the empirical evidence they assessed, Barker *et al.* (2007) concluded that the competitive losses of then-current mitigation actions, e.g., the EU ETS, were not significant.

Under the EU ETS rules Carbon Leakage Exposure Factor is used to determine the volumes of free allocation of emission permits to industrial installations.

Trade

To understand carbon trading, it is important to understand the products that are being traded. The primary product in carbon markets is the trading of GHG emission permits. Under a cap-and-trade system, permits are issued to various entities for the right to emit GHG emissions that meet emission reduction requirement caps.

One of the controversies about carbon mitigation policy is how to "level the playing field" with border adjustments. For example, one component of the American Clean Energy and Security Act (a 2009 bill that did not pass), along with several other energy bills put before US Congress, calls for carbon surcharges on goods imported from countries without cap-and-trade programs. Besides issues of compliance with the General Agreement on Tariffs and Trade, such border adjustments presume that the producing countries bear responsibility for the carbon emissions.

A general perception among developing countries is that discussion of climate change in trade negotiations could lead to "green protectionism" by high-income countries (World Bank, 2010, p. 251). Tariffs on imports ("virtual carbon") consistent with a carbon price of $50 per ton of CO_2 could be significant for developing countries. World Bank (2010) commented that introducing border tariffs could lead to a proliferation of trade measures where the competitive playing field is viewed as being uneven. Tariffs could also be a burden on low-income countries that have contributed very little to the problem of climate change.

Dust Collection System

A dust collection system is an air quality improvement system used in industrial, commercial, and home production shops to improve breathable air quality and safety by removing particulate matter from the air and environment. Dust collection systems work on the basic formula of *capture, convey* and *collect*.

First, the dust must be *captured*. This is accomplished with devices such as capture hoods to catch dust at its source of origin. Many times, the machine producing the dust will have a port to which a duct can be directly attached.

Second, the dust must be *conveyed*. This is done via a ducting system, properly sized and manifolded to maintain a consistent minimum air velocity required to keep the dust in suspension for conveyance to the collection device. A duct of the wrong size can lead to material settling in the duct system and clogging it.

Finally, the dust is *collected*. This is done via a variety of means, depending on the application and the dust being handled. It can be as simple as a basic pass-through filter, a cyclonic separator, or an impingement baffle. It can also be as complex as an electrostatic precipitator, a multistage baghouse, or a chemically treated wet scrubber or stripping tower.

Types of Systems

Smaller dust collection systems use a single-stage vacuum unit to create suction and perform air filtration, where the waste material is drawn into an impeller and deposited into a container such as a bag, barrel, or canister. Air is recirculated into the shop after passing through a filter to trap smaller particulate.

Larger systems utilize a two-stage system, which separates larger particles from fine dust using a pre-collection device, such as a cyclone or baffled canister, before drawing the air through the impeller. Air from these units can then be exhausted outdoors or filtered and recirculated back into the work space.

Dust collection systems are often part of a larger air quality management program that also includes large airborne particle filtration units mounted to the ceiling of shop spaces and mask systems to be worn by workers. Air filtration units are designed to process large volumes of air to remove fine particles (2 to 10 micrometres) suspended in the air. Masks are available in a variety of forms, from simple cotton face masks to elaborate respirators with tanked air — the need for which is determined by the environment in which the worker is operating.

In industry, round or rectangular ducts are used to prevent buildup of dust in processing equipment.

Dangers of Neglect

Proper dust collection and air filtration is important in any work space. Repeated exposure to wood dust can cause chronic bronchitis, emphysema, "flu-like" symptoms, and cancer. Wood dust also frequently contains chemicals and fungi, which can become airborne and lodge deeply in the lungs, causing illness and damage.

Carbon Dioxide Removal

These strategies deal with the root causes of climate change and involve reducing the amount of carbon dioxide already in the atmosphere (different from reducing the amount of carbon dioxide emitted). One method involves capturing carbon dioxide from ambient air, compressing it and storing it in geologic reservoirs (usually porous rock or depleted oil wells). However, at this point in time this strategy is considered too expensive to be cost effective.

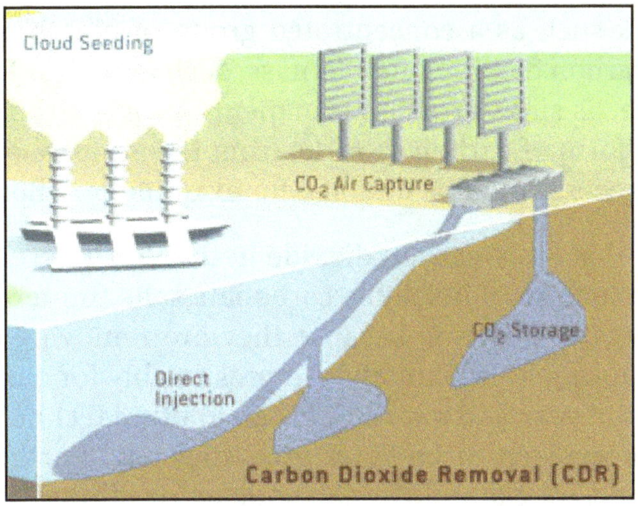

Another strategy, known as Ocean Iron Fertilization, involves adding iron to the ocean in order to increase algal blooms. These blooms consume carbon dioxide as they grow, and when they die, they sink to the ocean floor, effectively sequestering the carbon. However, this method is highly controversial, because it is not known how long the carbon could actually be stored for, or the negative impacts it could have on the ocean ecosystem (2). An approach similar to this involves adding minerals like calcium or magnesium oxide to the ocean. These minerals react with acidic carbon dioxide, and increase the ocean's capacity to hold and store carbon. Another benefit to this method is that it also helps to reduce ocean acidification. However, as with all CDR strategies, there has been little research into long term effects.

CDR strategies are generally not the first ones considered by those who are interested in the use of geoengineering. These strategies are thought of as needing to occur on a very large scale over a long period of time in order to actually be effective. For this reason, they tend to be very expensive and not a viable option at this time.

General

Carbon dioxide removal is different from reducing emissions, as the former produces an outlet of carbon dioxide from Earth's atmosphere, whereas the latter decreases the inlet of carbon dioxide to the atmosphere. Both have the same net effect, but for achieving carbon dioxide concentration levels below present levels, carbon dioxide removal is critical. Also for meeting higher concentration levels, carbon dioxide removal is increasingly considered to be crucial as it provides the only possibility to fill the gap between needed reductions to meet mitigation targets and global emission trends.

In the *OECD Environmental Outlook to 2050* released at the 2011 United Nations Climate Change Conference, the authors commented on the need for negative emissions, stating "Achieving lower concentration targets (450 ppm) depends significantly on the use of BECCS".

A carbon dioxide sink such as a concentrated group of plants or any other primary producer that binds carbon dioxide into biomass, such as within forests and kelp beds, is not carbon negative, as sinks are not permanent. A carbon dioxide sink of this type moves carbon, in the form of carbon dioxide, from the atmosphere or hydrosphere to the biosphere. This process could be undone, for example by wildfires or logging.

Carbon dioxide sinks that store carbon dioxide in the Earth's crust by injecting it into the subsurface, or in the form of insoluble carbonate salts (mineral sequestration), are considered carbon negative. This is because they are removing carbon from the atmosphere and sequestering it indefinitely and presumably for a considerable duration (thousands to millions of years). However, Carbon Capture technology remains, at best, theoretical and is yet to reach more than 33% efficiency. Furthermore, this process could be rapidly undone, for example by earthquakes or mining.

Methods

Bio-energy with Carbon Capture and Storage

Bio-energy with carbon capture and storage, or BECCS, uses biomass to extract carbon dioxide from the atmosphere, and carbon capture and storage technologies to concentrate and permanently store it in deep geological formations.

BECCS is currently (as of October 2012) the only CDR technology deployed at full industrial scale, with 550 000 tonnes CO_2/year in total capacity operating, divided between three different facilities (as of January 2012).

The Imperial College London, the UK Met Office Hadley Centre for Climate Prediction and Research, the Tyndall Centre for Climate Change Research, the Walker Institute for Climate System Research, and the Grantham Institute for Climate Change issued a joint report on carbon dioxide removal technologies as part of the *AVOID: Avoiding dangerous climate change* research program, stating that "Overall, of the technologies studied in this report, BECCS has the greatest maturity and there are no major practical barriers to its introduction into today's energy system. The presence of a primary product will support early deployment."

According to the OECD, "Achieving lower concentration targets (450 ppm) depends significantly on the use of BECCS".

Biochar

Biochar is created by the pyrolysis of biomass, and is under investigation as a method of carbon sequestration. Biochar is a charcoal that is used for agricultural purposes which also aids in carbon sequestration, the capture or hold of carbon. It is created using a process called pyrolysis, which is basically the act of high temperature heating biomass in an environment with low oxygen levels. What remains is a material known as char, similar to charcoal but is made through a sustainable process, thus the use of biomass. Biomass is organic matter produced by living organisms or recently living organisms, most commonly plants or plant based material. The offset of GHG emission, if biochar were to be implemented, would be a maximum of 12%. This equates to about 106 metric tons of CO_2 equivalents. On a medium conservative level, it would be 23% less than that, at 82 metric tons. A study done by the UK Biochar Research Center has stated that, on a conservative level, biochar can store 1 gigaton of carbon per year. With greater effort in marketing and acceptance of biochar, the benefit could be the storage of 5–9 gigatons per year of carbon in biochar soils.

Enhanced Weathering

Enhanced weathering refers to chemical approach to remove carbon dioxide involving land or ocean-based techniques. Examples of land based enhanced weathering techniques are in-situ carbonation of silicates. Ultramafic rock, for example, has the potential to store from hundreds to thousands of years worth of CO_2 emissions according to estimates. Ocean-based techniques involve alkalinity enhancement, such as, grinding, dispersing and dissolving olivine, limestone, silicates, or calcium hydroxide to address ocean acidification and CO_2 sequestration. Enhanced weathering is considered as one of the least expensive of geoengineering options. One example of a research project on the feasibility of enhanced weathering is the CarbFix project in Iceland.

Direct Air Capture (DAC)

Carbon dioxide can be removed from ambient air through chemical processes, sequestered, and stored. Traditional modes of carbon capture such as precombustion and postcombustion CO_2 capture from large point sources can help slow the rate of increase of the atmospheric CO_2 concentration, but only the direct removal of CO_2 from the air, or "direct air capture" (DAC), can actually reduce the global atmospheric CO_2 concentration if combined with long-term storage of CO_2.

A few engineering proposals have been made for removing CO_2 from the atmosphere, but work in this area is still in its infancy. Among the main technologies proposed, three of them stand out: Causticization with alkali and alkali-earth hydroxides, carbonation, and organic–inorganic hybrid sorbents consisting of amines supported in porous adsorbents.

One proposed method is by so-called *artificial trees*. This concept, proposed by climate scientist Wallace S. Broecker and science writer Robert Kunzig, imagines huge numbers of artificial trees around the world to remove ambient CO_2. The technology is now being pioneered by Klaus Lackner, a researcher at the Earth Institute, Columbia

University, whose artificial tree technology can suck up to 1,000 times more CO_2 from the air than real trees can, at a rate of about one ton of carbon per day if the artificial tree is approximately the size of an actual tree. The CO_2 would be captured in a filter and then removed from the filter and stored.

The chemistry used is a variant of that described below, as it is based on sodium hydroxide. However, in a more recent design proposed by Klaus Lackner, the process can be carried out at only 40 °C by using a polymer-based ion exchange resin, which takes advantage of changes in humidity to prompt the release of captured CO_2, instead of using a kiln. This reduces the energy required to operate the process.

Another substance which can be used are Metal-organic frameworks (or MOF's). A special MOF has been made specifically for locking CO_2 by Joeri Denayer.

In 2008, the Discovery Channel covered the work of David Keith, of University of Calgary, who built a tower, 4 feet wide and 20 feet tall (1.2×6.1 meters), with a fan at the bottom that sucks air in, which comes out again at the top. In the process, about half the CO_2 is removed from the air.

This device uses the chemical process described in detail below. The system demonstrated on the Discovery Channel was a 1/90,000th scale test system of the capture section; the reagents are regenerated in a separate facility. The main costs of a full plant will be the cost to build it, and the energy input to regenerate the chemicals and produce a pure stream of CO_2.

To put this into perspective, people in the U.S. emit about 20 tonnes of CO_2 per person annually. In other words, each person in the U.S. would require a tower like the one featured by the Discovery Channel to remove this amount of CO_2 from the air, requiring an annual 2 megawatt-hours of electricity to operate it. By comparison, a refrigerator consumes about 1.2 megawatt-hours annually (2001 figures). But, by combining many small systems such as this into one large system, the construction costs and energy use can be reduced.

It has been proposed that the Solar updraft tower to generate electricity from thermal air currents also be used at the same time for amine gravity scrubbing of CO_2. Some heat would be required to regenerate the amine.

A similar CO_2 scrubber has also been built by Carbon Engineering. Besides simply focusing on capturing the CO_2, the company also puts emphasis on reuse of the CO_2, for example in the production of fuels, which would thus be carbon-neutral.

Direct air capture has been proposed as a way of generating carbon-neutral organic chemicals, by harvesting the atmospheric compounds and then using them in the production and synthesis of polymers and fuels.

The Swiss-based Climeworks built the first industrial scale direct air capturing plant in Hinwil, in the canton of Zurich, Switzerland. It started up in May 2017, and could scrub up to 900 metric tons of CO_2 per year using heat from a local waste incineration plant. The CO2 was then pumped into local greenhouses, where it was used to help grow vegetables like tomatoes and cucumbers.

Climeworks and Reykjavik Energy also started up a small test direct air capturing plant in Hellisheidi, Iceland in October 2017. Using waste heat from a local geothermal power plant, the test plant captured up to 50 tons of CO_2 per year which was bound with water and injected over 700 metres underground into basaltic bedrock. The CO_2 reacted with the basalt to form solid carbonate minerals.

Ocean Fertilization

Ocean fertilization or ocean nourishment is a type of climate engineering based on the purposeful introduction of nutrients to the upper ocean to increase marine food production and to remove carbon dioxide from the atmosphere. A number of techniques, including fertilization by iron, urea and phosphorus have been proposed.

Diesel Particulate Filters

Diesel particulate filters (DPF) are devices that physically capture diesel particulates to prevent their release to the atmosphere. Diesel particulate filter materials have been developed that show impressive filtration efficiencies, in excess of 90%, as well as good mechanical and thermal durability. Diesel particulate filters have become the most effective technology for the control of diesel particulate emissions—including particle mass and numbers—with high efficiencies.

Due to the particle deposition mechanisms in these devices, filters are most effective in controlling the solid fraction of diesel particulates, including elemental carbon (soot) and the related black smoke emission. Filters may have limited effectiveness, or be totally ineffective, in controlling non-solid fractions of PM emissions—SOF and sulfate particulates. To control total PM emissions, DPF systems are likely to incorporate additional functional components targeting the SOF—typically oxidation catalysts—while ultra low sulfur fuels may be required to control sulfate particulates.

The term "diesel particulate trap" is sometimes used as a synonym for "diesel particulate filter", especially in older literature. The term "trap" covers a wider class of particle separation devices. Several particle deposition mechanisms other than filtration are commonly employed in industrial dust separation equipment. Examples include gravity settling, centrifugal separation, or electrostatic trapping. None of these techniques could be adopted to control diesel PM emissions, due to the small particle size and low density of diesel soot.

It may be noted that *particle oxidation catalysts* (POC)—sometimes called *partial filters*—can also capture diesel particulates, but provide a much lower overall efficiency than diesel particulate filters. In their common designs, POCs capture particulates only from a fraction of the flow, whereas the total flow is filtered in diesel particulate filters. In the case of some filter media, however, the distinction may not be very clear and the devices can be classified as either a POC or a (depth) particulate filter.

Collection & Regeneration: Due to the low bulk density of diesel particulates, which is typically below 0.1 g/cm³ (the density depends on the degree of compactness; as an example, a number of 0.056 g/cm³ was reported by Wade [Wade 1981]), diesel particulate filters can quickly accumulate considerable volumes of soot. Several liters of soot per day may be collected from an older generation heavy-duty truck or bus engine. The collected particulates would eventually cause excessively high exhaust gas pressure drop in the filter, which would negatively affect the engine operation. Therefore, diesel particulate filter systems have to provide a way of removing particulates from the filter to restore its soot collection capacity. This removal of particulates, known as the filter *regeneration*, can be performed either continuously, during regular operation of the filter, or periodically, after a pre-determined quantity of soot has been accumulated. In either case, the regeneration of filter systems should be "invisible" to the vehicle driver/operator and should be performed without his intervention. Thermal regeneration of diesel particulate filters is typically employed, where the collected particulates are oxidized—by oxygen and/or nitrogen dioxide—to gaseous products, primarily to carbon dioxide. Thermal regeneration, schematically represented in Figure, is undoubtedly the cleanest and most attractive method of operating diesel particulate filters.

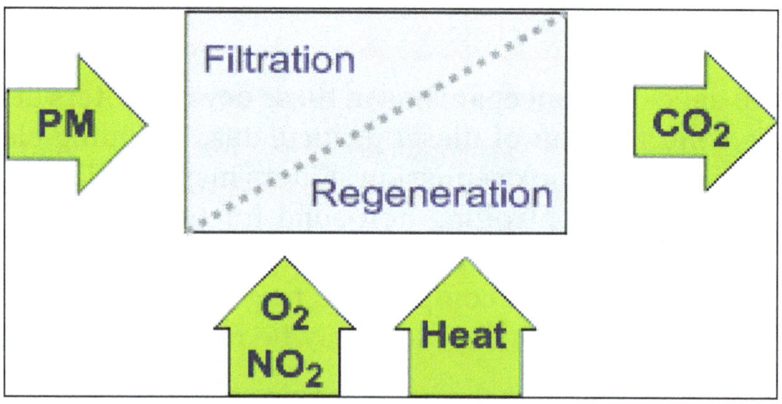

Schematic of particulate filter with thermal regeneration.

To ensure that particulates are oxidized at a sufficient rate, the filter must operate at a sufficient temperature. In some filter systems, the source of heat is the exhaust gas stream itself. In this type of filter system, referred to as a *passive filter*, the filter regenerates continuously during the regular operation of the engine. Passive filters usually incorporate some form of a catalyst, which lowers the soot oxidation temperature to a level that can be reached by exhaust gases during the operation of the vehicle.

Another approach which may be needed to facilitate reliable regeneration involves a number of active strategies for increasing the filter temperature (engine management, fuel combustion in the exhaust system, electric heaters, etc.). Regeneration of such devices, known as *active filters*, is usually performed periodically, as determined by the vehicle's control system. The classification of particulate filter systems is presented in more detail in the paper on filter systems.

An alternative strategy involves the use of disposable filter cartridges, which are replaced with new units once filled with soot. Particulate filters of this kind are used in some occupational health environments. Such maintenance intensive filter systems are clearly not acceptable in highway vehicle applications.

Mode of Action

Wall-flow diesel particulate filters usually remove 85% or more of the soot, and under certain conditions can attain soot removal efficiencies approaching 100%. Some filters are single-use, intended for disposal and replacement once full of accumulated ash. Others are designed to burn off the accumulated particulate either passively through the use of a catalyst or by active means such as a fuel burner which heats the filter to soot combustion temperatures. This is accomplished by engine programming to run (when the filter is full) in a manner that elevates exhaust temperature, in conjunction with an extra fuel injector in the exhaust stream that injects fuel to react with a catalyst element to burn off accumulated soot in the DPF filter, or through other methods. This is known as "filter regeneration". Cleaning is also required as part of periodic maintenance, and it must be done carefully to avoid damaging the filter. Failure of fuel injectors or turbochargers resulting in contamination of the filter with raw diesel or engine oil can also necessitate cleaning. The regeneration process occurs at road speeds higher than can generally be attained on city streets; vehicles driven exclusively at low speeds in urban traffic can require periodic trips at higher speeds to clean out the DPF. If the driver ignores the warning light and waits too long to operate the vehicle above 40 miles per hour (64 km/h), the DPF may not regenerate properly, and continued operation past that point may spoil the DPF completely so it must be replaced. Some newer diesel engines, namely those installed in combination vehicles, can also perform what is called a Parked Regeneration, where the engine increases RPM to around 1400 while parked, to increase the temperature of the exhaust.

Diesel engines produce a variety of particles during combustion of the fuel/air mix due to incomplete combustion. The composition of the particles varies widely dependent upon engine type, age, and the emissions specification that the engine was designed to meet. Two-stroke diesel engines produce more particulate per unit of power than do four-strokediesel engines, as they burn the fuel-air mix less completely.

Diesel particulate matter resulting from the incomplete combustion of diesel fuel produces soot (black carbon) particles. These particles include tiny nanoparticles—

smaller than a thousandth of a millimeter (one micron). Soot and other particles from diesel engines worsen the particulate matter pollution in the air and are harmful to health.

New particulate filters can capture from 30% to greater than 95% of the harmful soot. With an optimal diesel particulate filter (DPF), soot emissions may be decreased to 0.001 g / km or less.

The quality of the fuel also influences the formation of these particles. For example, a high sulfur content diesel produces more particles. Lower sulfur fuel produces fewer particles, and allows use of particulate filters. The injection pressure of diesel also influences the formation of fine particles.

DPF and NOx emissions strategies greatly increased fuel consumption in 2007 model year diesel engines, the addition of DEF fluid has reduced fuel consumption, but fuel consumption is still higher than in pre-emissions engines.

Variants of DPFs

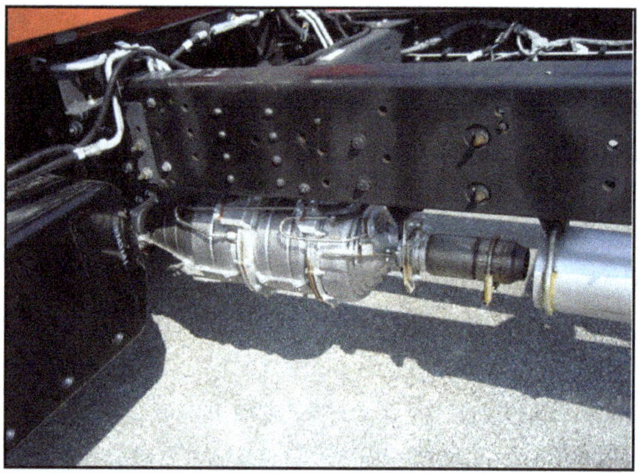

Cordierite Diesel Particulate Filter on GM 7.8 Isuzu.

Unlike a catalytic converter which is a flow-through device, a DPF retains bigger exhaust gas particles by forcing the gas to flow through the filter; however, the DPF does not retain small particles and maintenance-free DPFs break larger particles into smaller ones. There are a variety of diesel particulate filter technologies on the market. Each is designed around similar requirements:

1. Fine filtration
2. Minimum pressure drop
3. Low cost
4. Mass production suitability
5. Product durability

Cordierite Wall Flow Filters

The most common filter is made of cordierite (a ceramic material that is also used as catalytic converter supports (cores)). Cordierite filters provide excellent filtration efficiency, are relatively inexpensive, and have thermal properties that make packaging them for installation in the vehicle simple. The major drawback is that cordierite has a relatively low melting point (about 1200 °C) and cordierite substrates have been known to melt during filter regeneration. This is mostly an issue if the filter has become loaded more heavily than usual, and is more of an issue with passive systems than with active systems, unless there is a system break down.

Cordierite filter cores look like catalytic converter cores that have had alternate channels plugged - the plugs force the exhaust gas flow through the wall and the particulate collects on the inlet face.

Silicon Carbide Wall Flow Filters

The second most popular filter material is silicon carbide, or SiC. It has a higher (2700 °C) melting point than cordierite, however it is not as stable thermally, making packaging an issue. Small SiC cores are made of single pieces, while larger cores are made in segments, which are separated by a special cement so that heat expansion of the core will be taken up by the cement, and not the package. SiC cores are usually more expensive than cordierite cores, however they are manufactured in similar sizes, and one can often be used to replace the other. Silicon carbide filter cores also look like catalytic converter cores that have had alternate channels plugged - again the plugs force the exhaust gas flow through the wall and the particulate collects on the inlet face.

The characteristics of the wall flow diesel Particulate filter substrate are as follows: broad band filtration (the diameters of the filtered particles are 0.2-150 μm); high filtration efficiency (can be up to 95%); high refractory; high mechanical properties; high boiling point.

Ceramic Fiber Filters

Fibrous ceramic filters are made from several different types of ceramic fibers that are mixed together to form a porous media. This media can be formed into almost any shape and can be customized to suit various applications. The porosity can be controlled in order to produce high flow, lower efficiency or high efficiency lower volume filtration. Fibrous filters have an advantage over wall flow design of producing lower back pressure. Ceramic wall-flow filters remove carbon particulates almost completely, including fine particulates less than 100 nanometers (nm) diameter with an efficiency of greater than 95% in mass and greater than 99% in number of particles over a wide range of engine operating conditions. Since the continuous flow of soot into the filter would eventually block it, it is necessary to 'regenerate' the filtration properties of the filter by burning-off the collected particulate on a regular basis. Soot particulates burn-off forms water and CO_2 in small quantity amounting to less than 0.05% of the CO_2 emitted by the engine.

Metal Fiber flow-through Filters

Some cores are made from metal fibers - generally the fibers are "woven" into a monolith. Such cores have the advantage that an electrical current can be passed through the monolith to heat the core for regeneration purposes, allowing the filter to regenerate at low exhaust temperatures and/or low exhaust flow rates. Metal fiber cores tend to be more expensive than cordierite or silicon carbide cores, and generally not interchangeable with them because of the electrical requirement.

Paper

Disposable paper cores are used in certain specialty applications, without a regeneration strategy. Coal mines are common users — the exhaust gas is usually first passed through a water trap to cool it, and then through the filter. Paper filters are also used when a diesel machine must be used indoors for short periods of time, such as on a forklift being used to install equipment inside a store.

Partial Filters

There are a variety of devices that produce over 50% particulate matter filtration, but less than 85%. Partial filters come in a variety of materials. The only commonality between them is that they produce more back pressure than a catalytic converter, and less than a diesel particulate filter. Partial filter technology is popular for retrofit.

Maintenance

Filters require more maintenance than catalytic converters. Ash, a byproduct of oil consumption from normal engine operation, builds up in the filter as it cannot be converted into a gas and pass through the walls of the filter. This increases the pressure before the filter. Warnings are given to the driver before filter restriction causes an issue with drive-ability or damage to the engine or filter develop. Regular filter maintenance is a necessity.

DPF filters go through a regeneration process which removes this soot and lowers the filter pressure. There are three types of regeneration: active, passive, and forced. Active regeneration happens while the vehicle is not in use and takes 10 minutes on average to complete. Passive regeneration takes place while driving using the heat of the exhaust. This works well for vehicles that drive longer distances with few stops compared to those that perform short trips with many starts and stops. If the filter develops too much pressure then the last type of regeneration must be used - a forced regeneration. This involves a garage using a computer program to run the car, initiating a regeneration of the DPF manually.

There are three ways to clean your truck by using three regenerating cycle types. If using the active regeneration, the process ends in about 10 minutes. The long distances truck drivers are well familiarized with the passive regeneration when the cycle once started can only be ended in a mechanic shop. The third regeneration cycle supposes a forced process and occurs when the car is run after a computer program.

Safety

In 2011, Ford recalled 37,400 F-Series trucks with diesel engines after fuel and oil leaks caused fires in the diesel particulate filters of the trucks. No injuries occurred before the recall, though one grass fire was started. A similar recall was issued for 2005-2007 Jaguar S-Type and XJ diesels, where large amounts of soot became trapped in the DPF. In affected vehicles, smoke and fire emanated from the vehicle underside, accompanied by flames from the rear of the exhaust. The heat from the fire could cause heating through the transmission tunnel to the interior, melting interior components and potentially causing interior fires.

Flue-gas Desulfurization

FGD is a set of technologies used to remove sulphurdioxide (SO2) from exhaust Øue gases of fossil-fuel power plants, as well as from the emissions of other SOx emitting processes. Common methods used in it are wet scrubbing method, Wet and Dry lime scrubbing method, Spray-dry scrubbing method, SNOX method, Dry sorbent injection method, etc.

For a typical coal-×red power station, FGD system may remove 90% or more of the SO2 in the flue gases. SO2 emissions are a primary contributor to acid rain and have been regulated by every industrialized nation in the world.

Flue Gases

Flue Gases is mixture of gases produced by combustion of fuel and other materials in power stations and various industrial plants and released via flue (ducts) in atmosphere. It largely contains oxides of nitrogen derived from combustion of air, sulphur oxides, carbon dioxide, carbon monoxide, water vapour, excess oxygen, particulate matter like soot.

FGD Chemistry

Basic Principles

Most FGD systems employ two stages: one for fly ash removal and the other for SO_2 removal. Attempts have been made to remove both the fly ash and SO_2 in one scrubbing vessel. However, these systems experienced severe maintenance problems and low removal efficiency. In wet scrubbing systems, the flue gas normally passes first through a fly ash removal device, either an electrostatic precipitator or a baghouse, and then into the SO_2-absorber. However, in dry injection or spray drying operations, the SO_2 is first reacted with the lime, and then the flue gas passes through a particulate control device.

Another important design consideration associated with wet FGD systems is that the flue gas exiting the absorber is saturated with water and still contains some SO_2. These gases are highly corrosive to any downstream equipment such as fans, ducts, and stacks. Two methods that may minimize corrosion are: (1) reheating the gases to above their dew point, or (2) using materials of construction and designs that allow equipment to withstand the corrosive conditions. Both alternatives are expensive. Engineers determine which method to use on a site-by-site basis.

Scrubbing with an Alkali Solid or Solution

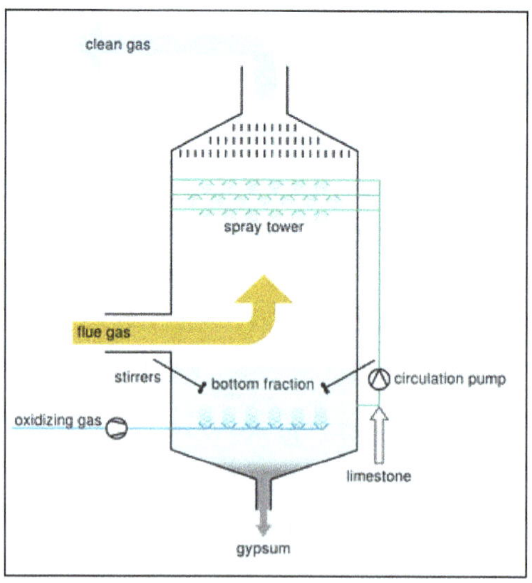

Schematic design of the absorber of an FGD.

SO_2 is an acid gas, and, therefore, the typical sorbent slurries or other materials used to remove the SO_2 from the flue gases are alkaline. The reaction taking place in wet scrubbing using a $CaCO_3$ (limestone) slurry produces calcium sulfite ($CaSO_3$) and may be expressed in the simplified dry form as:

$$CaCO_{3(s)} + SO_{2(g)} \rightarrow CaSO_{3(s)} + CO_{2(g)}$$

When wet scrubbing with a $Ca(OH)_2$ (hydrated lime) slurry, the reaction also produces $CaSO_3$ (calcium sulfite) and may be expressed in the simplified dry form as:

$$Ca(OH)_{2(s)} + SO_{2(g)} \rightarrow CaSO_{3(s)} + H_2O_{(l)}$$

When wet scrubbing with a $Mg(OH)_2$ (magnesium hydroxide) slurry, the reaction produces $MgSO_3$ (magnesium sulfite) and may be expressed in the simplified dry form as:

$$Mg(OH)_{2(s)} + SO_{2(g)} \rightarrow MgSO_{3(s)} + H_2O_{(l)}$$

To partially offset the cost of the FGD installation, some designs, particularly dry sorbent injection systems, further oxidize the $CaSO_3$ (calcium sulfite) to produce marketable $CaSO_4$-$2H_2O$ (gypsum) that can be of high enough quality to use in wallboardand other products. The process by which this synthetic gypsum is created is also known as forced oxidation:

$$CaSO_{3(aq)} + 2H_2O_{(l)} + ½O_{2(g)} \rightarrow CaSO_4 \cdot 2H_2O_{(s)}$$

A natural alkaline usable to absorb SO_2 is seawater. The SO_2 is absorbed in the water, and when oxygen is added reacts to form sulfate ions SO_4- and free H^+. The surplus of H^+ is offset by the carbonates in seawater pushing the carbonate equilibrium to release CO_2 gas:

$$SO_{2(g)} + H_2O_{(l)} + ½O_{2(g)} \rightarrow SO_4^{2-}{}_{(aq)} + 2H^+$$

$$HCO_3^- + H^+ \rightarrow H_2O_{(l)} + CO_{2(g)}$$

In industry caustic (NaOH) is often used to scrub SO_2, producing sodium sulfite:

$$2NaOH_{(aq)} + SO_{2(g)} \rightarrow Na_2SO_{3(aq)} + H_2O_{(l)}$$

Types of Wet Scrubbers used in FGD

To promote maximum gas–liquid surface area and residence time, a number of wet scrubber designs have been used, including spray towers, venturis, plate towers, and mobile packed beds. Because of scale buildup, plugging, or erosion, which affect FGD dependability and absorber efficiency, the trend is to use simple scrubbers such as spray towers instead of more complicated ones. The configuration of the tower may be vertical or horizontal, and flue gas can flow cocurrently, countercurrently, or crosscurrently with respect to the liquid. The chief drawback of spray towers is that they require a higher liquid-to-gas ratio requirement for equivalent SO_2 removal than other absorber designs.

FGD scrubbers produce a scaling wastewater that requires treatment to meet discharge regulations. However, technological advancements in ion exchange membranes and el ectrodialysis systems has enabled high-efficiency treatment of FGD wastewater to meet recent EPA discharge limits. The treatment approach is similar for other highly scaling industrial wastewaters.

Venturi Scrubbers

A venturi scrubber is a converging/diverging section of duct. The converging section accelerates the gas stream to high velocity. When the liquid stream is injected at the throat, which is the point of maximum velocity, the turbulence caused by the high gas velocity atomizes the liquid into small droplets, which creates the surface area necessary for mass transfer to take place. The higher the pressure drop in the venturi, the smaller the droplets and the higher the surface area. The penalty is in power consumption.

For simultaneous removal of SO_2 and fly ash, venturi scrubbers can be used. In fact, many of the industrial sodium-based throwaway systems are venturi scrubbers originally designed to remove particulate matter. These units were slightly modified to inject a sodium-based scrubbing liquor. Although removal of both particles and SO_2 in one vessel can be economic, the problems of high pressure drops and finding a scrubbing medium to remove heavy loadings of fly ash must be considered. However, in cases where the particle concentration is low, such as from oil-fired units, it can be more effective to remove particulate and SO_2 simultaneously.

Packed Bed Scrubbers

A packed scrubber consists of a tower with packing material inside. This packing material can be in the shape of saddles, rings, or some highly specialized shapes designed to maximize the contact area between the dirty gas and liquid. Packed towers typically operate at much lower pressure drops than venturi scrubbers and are therefore cheaper to operate. They also typically offer higher SO_2 removal efficiency. The drawback is that they have a greater tendency to plug up if particles are present in excess in the exhaust air stream.

Spray Towers

A spray tower is the simplest type of scrubber. It consists of a tower with spray nozzles, which generate the droplets for surface contact. Spray towers are typically used when circulating a slurry. The high speed of a venturi would cause erosion problems, while a packed tower would plug up if it tried to circulate a slurry.

Counter-current packed towers are infrequently used because they have a tendency to become plugged by collected particles or to scale when lime or limestone scrubbing slurries are used.

Scrubbing Reagent

Alkaline sorbents are used for scrubbing flue gases to remove SO_2. Depending on the application, the two most important are lime and sodium hydroxide (also known as caustic soda). Lime is typically used on large coal- or oil-fired boilers as found in power plants, as it is very much less expensive than caustic soda. The problem is that it results in a slurry being circulated through the scrubber instead of a solution. This makes it harder on the equipment. A spray tower is typically used for this application. The use of lime results in a slurry of calcium sulfite ($CaSO_3$) that must be disposed of. Fortunately, calcium sulfite can be oxidized to produce by-product gypsum ($CaSO_4 \cdot 2H_2O$) which is marketable for use in the building products industry.

Caustic soda is limited to smaller combustion units because it is more expensive than lime, but it has the advantage that it forms a solution rather than a slurry. This makes it easier to operate. It produces a "spent caustic" solution of sodium sulfite/bisulfite (depending on the pH), or sodium sulfate that must be disposed of. This is not a problem in a kraft pulp mill for example, where this can be a source of makeup chemicals to the recovery cycle.

Scrubbing with Sodium Sulfite Solution

It is possible to scrub sulfur dioxide by using a cold solution of sodium sulfite; this forms a sodium hydrogen sulfite solution. By heating this solution it is possible to reverse the reaction to form sulfur dioxide and the sodium sulfite solution. Since the sodium sulfite solution is not consumed, it is called a regenerative treatment. The application of this reaction is also known as the Wellman–Lord process.

In some ways this can be thought of as being similar to the reversible liquid–liquid extraction of an inert gas such as xenon or radon (or some other solute which does not undergo a chemical change during the extraction) from water to another phase. While a chemical change does occur during the extraction of the sulfur dioxide from the gas mixture, it is the case that the extraction equilibrium is shifted by changing the temperature rather than by the use of a chemical reagent.

Gas Phase Oxidation Followed by Reaction with Ammonia

A new, emerging flue gas desulfurization technology has been described by the IAEA. It is a radiation technology where an intense beam of electrons is fired into the flue gas at the same time as ammonia is added to the gas. The Chendu power plant in China started up such a flue gas desulfurization unit on a 100 MW scale in 1998. The Pomorzany power plant in Poland also started up a similar sized unit in 2003 and that plant removes both sulfur and nitrogen oxides. Both plants are reported to be operating successfully. However, the accelerator design principles and manufacturing quality need further improvement for continuous operation in industrial conditions.

No radioactivity is required or created in the process. The electron beam is generated by a device similar to the electron gun in a TV set. This device is called an accelerator. This is an example of a radiation chemistry process where the physical effects of radiation are used to process a substance.

The action of the electron beam is to promote the oxidation of sulfur dioxide to sulfur(VI) compounds. The ammonia reacts with the sulfur compounds thus formed to produce ammonium sulfate, which can be used as a nitrogenous fertilizer. In addition, it can be used to lower the nitrogen oxide content of the flue gas. This method has attained industrial plant scale.

Dry Lime Scrubbing

In dry scrubbing, lime is injected directly into flue gas to remove SO_2 and HCl. There are two major dry processes: "dry injection" systems inject dry hydrated lime into the flue gas duct and "spray dryers" inject an atomized lime slurry into a separate vessel. A spray dryer is typically shaped like a silo, with a cylindrical top and a cone bottom. Hot flue gas flows into the top. Lime slurry is sprayed through an atomizer (*e.g.*, nozzles) into the cylinder near the top, where it absorbs SO_2 and HCl. The water in the lime slurry is then evaporated by the hot gas. The scrubbed flue gas flows from the bottom of the cylindrical section through a horizontal duct. A portion of the dried, unreacted lime and its reaction products fall to the bottom of the cone and are removed. The flue gas then flows to a particulate control device (*e.g.*, a baghouse) to remove the remainder of the lime and reaction products. Both dry injection and spray dryers yield a dry final product, collected in particulate control devices. At electric generating plants, dry scrubbing is used primarily for low-sulfur fuels. At municipal waste-to-energy plants, dry scrubbing is used for removal of SO_2 and HCl. Dry scrubbing is used at other industrial facilities for HCl control. Dry scrubbing methods have improved significantly in recent years, resulting in excellent removal efficiencies.

Wet Lime Scrubbing

In wet lime scrubbing, lime is added to water and the resulting slurry is sprayed into a flue gas scrubber. In a typical system, the gas to be cleaned enters the bottom of a cylinder-like tower and flows upward through a shower of lime slurry. The sulfur dioxide is absorbed into the spray and then precipitated as wet calcium sulfite. The sulfite can be converted to gypsum, a salable by-product. Wet scrubbing treats high-sulfur fuels and some low-sulfur fuels where high-efficiency sulfur dioxide removal is required. Wet scrubbing primarily uses magnesium-enhanced lime (containing 3-8% magnesium oxide) because it provides high alkalinity to increase SO_2 removal capacity and reduce scaling potential.

Comparing Lime and Limestone SO_2 Wet Scrubbing Processes

More than ninety percent of U.S. flue gas desulfurization (FGD) system capacity uses lime or limestone. This trend will likely continue into the next phase of federally mandated SO_2 reduction from coal burning power plants. In 2003, the National Lime Association sponsored a study by Sargent and Lundy to compare the costs of leading lime and limestone-based FGD processes utilized by power generating plants in the United States. The study included developing conceptual designs with capital and O&M

cost requirements using up-to-date performance criteria for the processes. The results of the study are summarized in two reports: Wet FGD Technology Evaluation and Dry FGD Technology Evaluation. The reports present the competitive position of wet and dry limestone and lime-based processes relative to reagent cost, auxiliary power cost, coal sulfur content, dispatch, capital cost, and by-product production (gypsum and SO_3 aerosol mitigation chemicals).

Cyclonic Separation

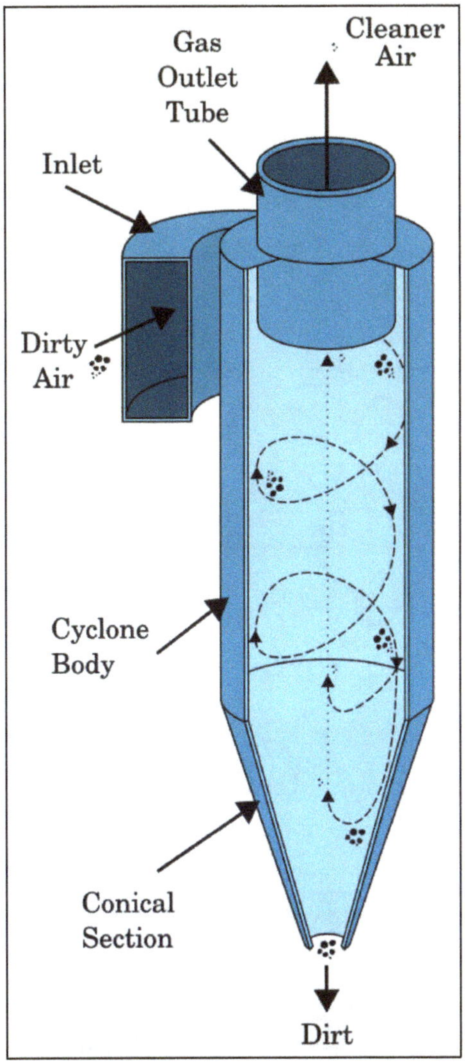

A simple cyclone separator.

Cyclonic separation is a method of removing particulates from an air, gas or liquid stream, without the use of filters, through vortex separation. When removing particulate matter from liquids, a hydrocyclone is used; while from gas, a gas cyclone is used. Rotational effects and gravity are used to separate mixtures of solids and fluids. The method can also be used to separate fine droplets of liquid from a gaseous stream.

A high speed rotating (air)flow is established within a cylindrical or conical container called a cyclone. Air flows in a helical pattern, beginning at the top (wide end) of the cyclone and ending at the bottom (narrow) end before exiting the cyclone in a straight stream through the center of the cyclone and out the top. Larger (denser) particles in the rotating stream have too much inertia to follow the tight curve of the stream, and strike the outside wall, then fall to the bottom of the cyclone where they can be removed. In a conical system, as the rotating flow moves towards the narrow end of the cyclone, the rotational radius of the stream is reduced, thus separating smaller and smaller particles. The cyclone geometry, together with flow rate, defines the *cut point* of the cyclone. This is the size of particle that will be removed from the stream with a 50% efficiency. Particles larger than the cut point will be removed with a greater efficiency, and smaller particles with a lower efficiency.

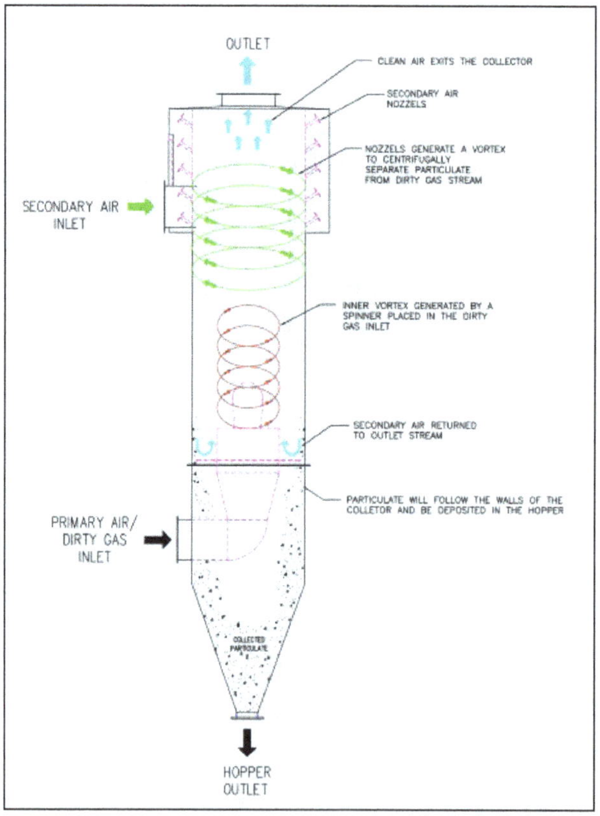

Airflow diagram for Aerodyne cyclone in standard vertical position. Secondary air flow is injected to reduce wall abrasion.

An alternative cyclone design uses a secondary air flow within the cyclone to keep the collected particles from striking the walls, to protect them from abrasion. The primary air flow containing the particulates enters from the bottom of the cyclone and is forced into spiral rotation by stationary spinner vanes. The secondary air flow enters from the top of the cyclone and moves downward toward the bottom, intercepting the particulate from the primary air. The secondary air flow also allows the collector to optionally be mounted horizontally, because it pushes the particulate toward the collection area, and does not rely solely on gravity to perform this function.

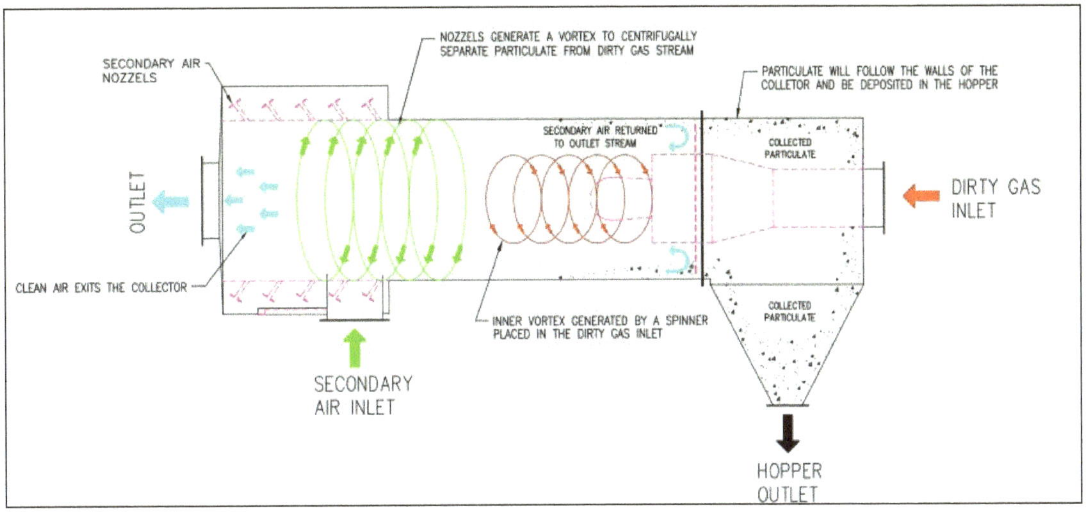

Airflow diagram for Aerodyne cyclone in horizontal position, an alternate design. Secondary air flow is injected to reduce wall abrasion, and to help move collected particulates to hopper for extraction.

Large scale cyclones are used in sawmills to remove sawdust from extracted air. Cyclones are also used in oil refineries to separate oils and gases, and in the cement industry as components of kiln preheaters. Cyclones are increasingly used in the household, as the core technology in bagless types of portable vacuum cleaners and central vacuum cleaners. Cyclones are also used in industrial and professional kitchen ventilation for separating the grease from the exhaust air in extraction hoods. Smaller cyclones are used to separate airborne particles for analysis. Some are small enough to be worn clipped to clothing, and are used to separate respirable particles for later analysis.

Similar separators are used in the oil refining industry (e.g. for Fluid catalytic cracking) to achieve fast separation of the catalyst particles from the reacting gases and vapors.

Analogous devices for separating particles or solids from liquids are called hydrocyclones or hydroclones. These may be used to separate solid waste from water in wastewater and sewage treatment.

Cyclone Theory

As the cyclone is essentially a two phase particle-fluid system, fluid mechanics and particle transport equations can be used to describe the behaviour of a cyclone. The air in a cyclone is initially introduced tangentially into the cyclone with an inlet velocity V_{in}. Assuming that the particle is spherical, a simple analysis to calculate critical separation particle sizes can be established.

If one considers an isolated particle circling in the upper cylindrical component of the cyclone at a rotational radius of r from the cyclone's central axis, the particle is therefore subjected to drag, centrifugal, and buoyant forces. Given that the fluid velocity is moving in a spiral the gas velocity can be broken into two component velocities: a tangential component, V_t, and an outward radial velocity component V_r. Assuming Stokes' law, the drag force in the outward radial direction that is opposing the outward velocity on any particle in the inlet stream is:

$$F_d = -6\pi r_p \mu V_r.$$

Using ρ_p as the particles density, the centrifugal component in the outward radial direction is:

$$F_c = m\frac{V_t^2}{r}$$

$$= \frac{4}{3}\pi\rho_p r_p^3 \frac{V_t^2}{r}.$$

The buoyant force component is in the inward radial direction. It is in the opposite direction to the particle's centrifugal force because it is on a volume of fluid that is missing compared to the surrounding fluid. Using ρ_f for the density of the fluid, the buoyant force is:

$$F_b = -V_p\rho_f \frac{V_t^2}{r}$$

$$= -\frac{4\pi r_p^3}{3}\frac{V_t^2}{r}\rho_f.$$

In this case, V_p is equal to the volume of the particle (as opposed to the velocity). Determining the outward radial motion of each particle is found by setting Newton's second law of motion equal to the sum of these forces:

$$m\frac{dV_r}{dt} = F_d + F_c + F_b$$

To simplify this, we can assume the particle under consideration has reached "terminal velocity", i.e., that its acceleration $\frac{dV_r}{dt}$ is zero. This occurs when the radial velocity has caused enough drag force to counter the centrifugal and buoyancy forces. This simplification changes our equation to:

$$F_d + F_c + F_b = 0$$

Which expands to:

$$-6\pi r_p \mu V_r + \frac{4}{3}\pi r_p^3 \frac{V_t^2}{r}\rho_p - \frac{4}{3}\pi r_p^3 \frac{V_t^2}{r}\rho_f = 0$$

Solving for V_r we have,

$$V_r = \frac{2}{9}\frac{r_p^2}{\mu}\frac{V_t^2}{r}(\rho_p - \rho_f).$$

Notice that if the density of the fluid is greater than the density of the particle, the motion is (-), toward the center of rotation and if the particle is denser than the fluid, the motion is (+), away from the center. In most cases, this solution is used as guidance in designing a separator, while actual performance is evaluated and modified empirically.

In non-equilibrium conditions when radial acceleration is not zero, the general equation from above must be solved. Rearranging terms we obtain

$$\frac{dV_r}{dt}+\frac{9}{2}\frac{\mu}{\rho_p r_p^2}V_r-\left(1-\frac{\rho_f}{\rho_p}\right)\frac{V_t^2}{r}=0$$

Since V_r is distance per time, this is a 2nd order differential equation of the form $x''+c_1x'+c_2=0$..

Experimentally it is found that the velocity component of rotational flow is proportional to r^2, therefore:

$$V_t \propto r^2.$$

This means that the established feed velocity controls the vortex rate inside the cyclone, and the velocity at an arbitrary radius is therefore:

$$Ur=U_{in}\frac{r}{R_{in}}.$$

Subsequently, given a value for V_t, possibly based upon the injection angle, and a cutoff radius, a characteristic particle filtering radius can be estimated, above which particles will be removed from the gas stream.

Alternative Models

The above equations are limited in many regards. For example, the geometry of the separator is not considered, the particles are assumed to achieve a steady state and the effect of the vortex inversion at the base of the cyclone is also ignored, all behaviours which are unlikely to be achieved in a cyclone at real operating conditions.

More complete models exist, as many authors have studied the behaviour of cyclone separators. Numerical modelling using computational fluid dynamics has also been used extensively in the study of cyclonic behaviour. A major limitation of any fluid mechanics model for cyclone separators is the inability to predict the agglomeration of fine particles with larger particles, which has a great impact on cyclone collection efficiency.

Electrostatic Precipitator

Electrodes inside electrostatic precipitator.

An electrostatic precipitator (ESP) is a filtration device that removes fine particles, like dust and smoke, from a flowing gas using the force of an induced electrostatic charge minimally impeding the flow of gases through the unit.

Collection electrode of electrostatic precipitator in waste incineration plant.

In contrast to wet scrubbers which apply energy directly to the flowing fluid medium, an ESP applies energy only to the particulate matter being collected and therefore is very efficient in its consumption of energy (in the form of electricity).

Invention of the Electrostatic Precipitator

The first use of corona discharge to remove particles from an aerosol was by Hohlfeld in 1824. However, it was not commercialized until almost a century later.

In 1907 Frederick Gardner Cottrell, a professor of chemistry at the University of California, Berkeley, applied for a patent on a device for charging particles and then collecting them through electrostatic attraction—the first electrostatic precipitator. Cottrell first applied the device to the collection of sulphuric acid mist and lead oxide fumes emitted from various acid-making and smelting activities. Wine-producing vineyards in northern California were being adversely affected by the lead emissions.

At the time of Cottrell's invention, the theoretical basis for operation was not understood. The operational theory was developed later in Germany, with the work of Walter Deutsch and the formation of the Lurgi company.

Cottrell used proceeds from his invention to fund scientific research through the creation of a foundation called Research Corporation in 1912, to which he assigned the patents. The intent of the organization was to bring inventions made by educators (such as Cottrell) into the commercial world for the benefit of society at large. The operation of Research Corporation is funded by royalties paid by commercial firms after commercialization occurs. Research Corporation has provided vital funding to many scientific projects: Goddard's rocketry experiments, Lawrence's cyclotron, production methods for vitamins A and B_1, among many others.

By a decision of the US Supreme Court, the Corporation had to be split into several entities. The Research Corporation was separated from two commercial firms making the hardware: Research-Cottrell Inc. (operating east of the Mississippi River) and Western Precipitation (operating in the western states). The Research Corporation continues to be active to this day, and the two companies formed to commercialize the invention for industrial and utility applications are still in business as well.

Electrophoresis is the term used for migration of gas-suspended charged particles in a direct-current electrostatic field. Traditional CRT television sets tend to accumulate dust on the screen because of this phenomenon (a CRT is a direct-current machine operating at about 35 kilovolts).

Plate Precipitator

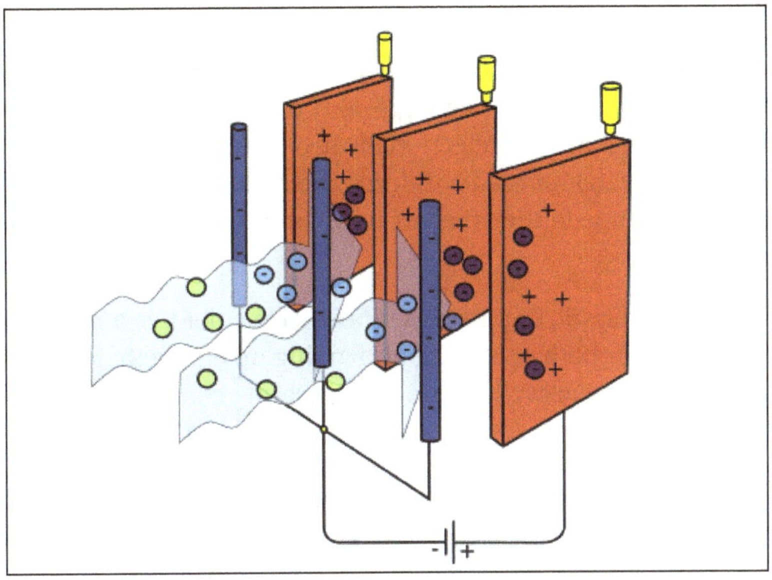

Conceptual diagram of an electrostatic precipitator.

The most basic precipitator contains a row of thin vertical wires, and followed by a stack of large flat metal plates oriented vertically, with the plates typically spaced about 1 cm to 18 cm apart, depending on the application. The air stream flows horizontally through the spaces between the wires, and then passes through the stack of plates.

A negative voltage of several thousand volts is applied between wire and plate. If the applied voltage is high enough, an electric corona discharge ionizes the air around the electrodes, which then ionizes the particles in the air stream.

The ionized particles, due to the electrostatic force, are diverted towards the grounded plates. Particles build up on the collection plates and are removed from the air stream.

A two-stage design (separate charging section ahead of collecting section) has the benefit of minimizing ozone production, which would adversely affect health of personnel working in enclosed spaces. For shipboard engine rooms where gearboxes generate an oil mist, two-stage ESP's are used to clean the air, improving the operating environment and preventing buildup of flammable oil fog accumulations. Collected oil is returned to the gear lubricating system.

Collection Efficiency (*R*)

Precipitator performance is very sensitive to two particulate properties: 1) Electrical resistivity; and 2) Particle size distribution. These properties can be measured economically and accurately in the laboratory, using standard tests. Resistivity can be determined as a function of temperature in accordance with IEEE Standard 548. This test is conducted in an air environment containing a specified moisture concentration. The test is run as a function of ascending or descending temperature, or both. Data is acquired using an average ash layer[further explanation needed] electric field of 4 kV/cm. Since relatively low applied voltage is used and no sulfuric acid vapor is present in the test environment, the values obtained indicate the maximum ash resistivity.

In an ESP, where particle charging and discharging are key functions, resistivity is an important factor that significantly affects collection efficiency. While resistivity is an important phenomenon in the inter-electrode region where most particle charging takes place, it has a particularly important effect on the dust layer at the collection electrode where discharging occurs. Particles that exhibit high resistivity are difficult to charge. But once charged, they do not readily give up their acquired charge on arrival at the collection electrode. On the other hand, particles with low resistivity easily become charged and readily release their charge to the grounded collection plate. Both extremes in resistivity impede the efficient functioning of ESPs. ESPs work best under normal resistivity conditions.

Resistivity, which is a characteristic of particles in an electric field, is a measure of a particle's resistance to transferring charge (both accepting and giving up charges). Resistivity is a function of a particle's chemical composition as well as flue gas operating conditions such as temperature and moisture. Particles can have high, moderate (normal), or low resistivity.

Bulk resistivity is defined using a more general version of Ohm's Law, as given in Equation (1) below:

$$\vec{E} = \rho \vec{j} \qquad (1)$$

Where:

E is the Electric field strength (V/cm);

j is the Current density (A/cm^2);

ρ is the Resistivity (Ohm-cm).

A better way of displaying this would be to solve for resistivity as a function of applied voltage and current, as given in Equation (2) below:

$$\rho = \frac{AV}{Il} \qquad (2)$$

Where:

ρ = Resistivity (Ohm-cm);

V = The applied DC potential, (Volts);

I = The measured current, (Amperes);

l = The ash layer thickness, (cm);

A = The current measuring electrode face area, (cm^2).

Resistivity is the electrical resistance of a dust sample 1.0 cm^2 in cross-sectional area, 1.0 cm thick, and is recorded in units of ohm-cm. A method for measuring resistivity will be described in this article. The table below, gives value ranges for low, normal, and high resistivity.

Resistivity	Range of Measurement
Low	between 10^4 and 10^7 ohm-cm
Normal	between 10^7 and 2×10^{10} ohm-cm
High	above 2×10^{10} ohm-cm

Dust Layer Resistivity

Resistivity affects electrical conditions in the dust layer by a potential electric field (voltage drop) being formed across the layer as negatively charged particles arrive at its surface and leak their electrical charges to the collection plate. At the metal surface of the electrically grounded collection plate, the voltage is zero, whereas at the outer surface of the dust layer, where new particles and ions are arriving, the electrostatic voltage caused by the gas ions can be quite high. The strength of this electric field depends on the resistivity and thickness of the dust layer.

In high-resistivity dust layers, the dust is not sufficiently conductive, so electrical charges have difficulty moving through the dust layer. Consequently, electrical charges accumulate on and beneath the dust layer surface, creating a strong electric field.

Voltages can be greater than 10,000 volts. Dust particles with high resistivities are held too strongly to the plate, making them difficult to remove and causing rapping problems.

In low resistivity dust layers, the corona current is readily passed to the grounded collection electrode. Therefore, a relatively weak electric field, of several thousand volts, is maintained across the dust layer. Collected dust particles with low resistivity do not adhere strongly enough to the collection plate. They are easily dislodged and become re-entrained in the gas stream.

The electrical conductivity of a bulk layer of particles depends on both surface and volume factors. Volume conduction, or the motions of electrical charges through the interiors of particles, depends mainly on the composition and temperature of the particles. In the higher temperature regions, above 500 °F (260 °C), volume conduction controls the conduction mechanism. Volume conduction also involves ancillary factors, such as compression of the particle layer, particle size and shape, and surface properties.

Volume conduction is represented in the figures as a straight-line at temperatures above 500 °F (260 °C). At temperatures below about 450 °F (230 °C), electrical charges begin to flow across surface moisture and chemical films adsorbed onto the particles. Surface conduction begins to lower the resistivity values and bend the curve downward at temperatures below 500 °F (260 °C).

These films usually differ both physically and chemically from the interiors of the particles owing to adsorption phenomena. Theoretical calculations indicate that moisture films only a few molecules thick are adequate to provide the desired surface conductivity. Surface conduction on particles is closely related to surface-leakage currents occurring on electrical insulators, which have been extensively studied. An interesting practical application of surface-leakage is the determination of dew point by measurement of the current between adjacent electrodes mounted on a glass surface. A sharp rise in current signals the formation of a moisture film on the glass. This method has been used effectively for determining the marked rise in dew point, which occurs when small amounts of sulfuric acid vapor are added to an atmosphere (commercial Dewpoint Meters are available on the market).

The following discussion of normal, high, and low resistivity applies to ESPs operated in a dry state; resistivity is not a problem in the operation of wet ESPs because of the moisture concentration in the ESP. The relationship between moisture content and resistivity is explained later in this work.

Normal Resistivity

As stated above, ESPs work best under normal resistivity conditions. Particles with normal resistivity do not rapidly lose their charge on arrival at the collection electrode. These particles slowly leak their charge to grounded plates and are retained on the collection plates by intermolecular adhesive and cohesive forces. This allows a particulate layer to be built up and then dislodged from the plates by rapping. Within the range of normal dust resistivity (between 10^7 and 2×10^{10} ohm-cm), fly ash is collected more easily than dust having either low or high resistivity.

High Resistivity

If the voltage drop across the dust layer becomes too high, several adverse effects can occur. First, the high voltage drop reduces the voltage difference between the discharge electrode and collection electrode, and thereby reduces the electrostatic field strength used to drive the gas ion-charged particles over to the collected dust layer. As the dust layer builds up, and the electrical charges accumulate on the surface of the dust layer, the voltage difference between the discharge and collection electrodes decreases. The migration velocities of small particles are especially affected by the reduced electric field strength.

Another problem that occurs with high resistivity dust layers is called back corona. This occurs when the potential drop across the dust layer is so great that corona discharges begin to appear in the gas that is trapped within the dust layer. The dust layer breaks down electrically, producing small holes or craters from which back corona discharges occur. Positive gas ions are generated within the dust layer and are accelerated toward the "negatively charged" discharge electrode. The positive ions reduce some of the negative charges on the dust layer and neutralize some of the negative ions on the "charged particles" heading toward the collection electrode. Disruptions of the normal corona process greatly reduce the ESP's collection efficiency, which in severe cases, may fall below 50% . When back corona is present, the dust particles build up on the electrodes forming a layer of insulation. Often this can not be repaired without bringing the unit offline.

The third, and generally most common problem with high resistivity dust is increased electrical sparking. When the sparking rate exceeds the "set spark rate limit," the automatic controllers limit the operating voltage of the field. This causes reduced particle charging and reduced migration velocities toward the collection electrode. High resistivity can generally be reduced by doing the following:

- Adjusting the temperature.
- Increasing moisture content.
- Adding conditioning agents to the gas stream.
- Increasing the collection surface area.
- Using hot-side precipitators (occasionally and with foreknowledge of sodium depletion).

Thin dust layers and high-resistivity dust especially favor the formation of back corona craters. Severe back corona has been observed with dust layers as thin as 0.1 mm, but a dust layer just over one particle thick can reduce the sparking voltage by 50%. The most marked effects of back corona on the current-voltage characteristics are:

1. Reduction of the spark over voltage by as much as 50% or more.
2. Current jumps or discontinuities caused by the formation of stable back-corona craters.
3. Large increase in maximum corona current, which just below spark over corona gap may be several times the normal current.

The Figure below and to the left shows the variation in resistivity with changing gas temperature for six different industrial dusts along with three coal-fired fly ashes. The Figure on the right illustrates resistivity values measured for various chemical compounds that were prepared in the laboratory.

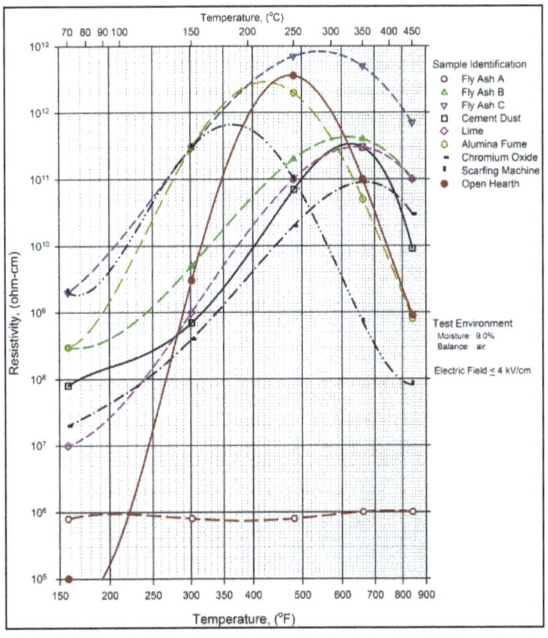

Resistivity Values of Representative Dusts and Fumes From Industrial Plants.

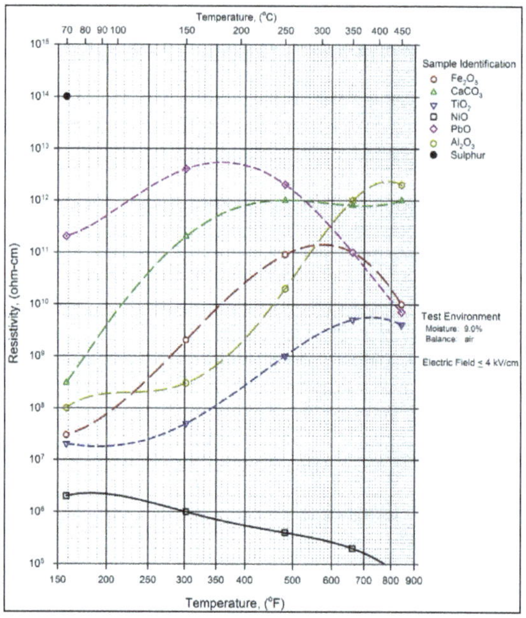

Resistivity Values of Various Chemicals and Reagents as a Function of Temperature.

Results for Fly Ash A (in the figure to the left) were acquired in the ascending temperature mode. These data are typical for a moderate to high combustibles content ash. Data for Fly Ash B are from the same sample, acquired during the descending temperature mode.

The differences between the ascending and descending temperature modes are due to the presence of unburned combustibles in the sample. Between the two test modes, the samples are equilibrated in dry air for 14 hours (overnight) at 850 °F (450 °C). This overnight annealing process typically removes between 60% and 90% of any unburned combustibles present in the samples. Exactly how carbon works as a charge carrier is not fully understood, but it is known to significantly reduce the resistivity of a dust.

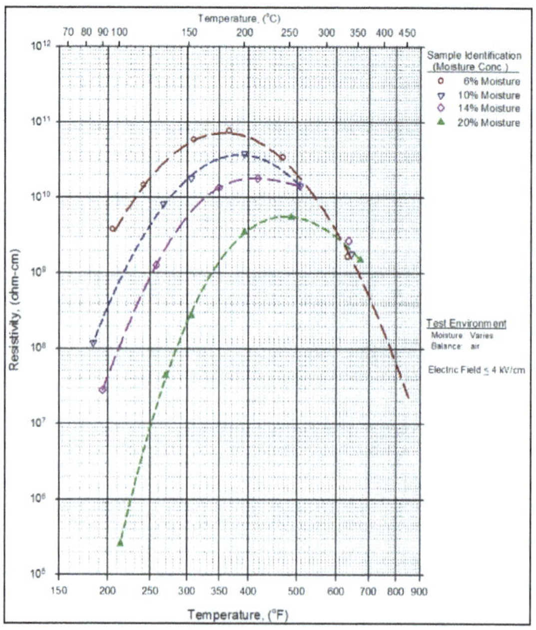

Resistivity Measured as a Function of Temperature in Varying Moisture Concentrations (Humidity).

Carbon can act, at first, like a high resistivity dust in the precipitator. Higher voltages can be required in order for corona generation to begin. These higher voltages can be problematic for the TR-Set controls. The problem lies in onset of corona causing large amounts of current to surge through the (low resistivity) dust layer. The controls sense this surge as a spark. As precipitators are operated in spark-limiting mode, power is terminated and the corona generation cycle re-initiates. Thus, lower power (current) readings are noted with relatively high voltage readings.

The same thing is believed to occur in laboratory measurements. Parallel plate geometry is used in laboratory measurements without corona generation. A stainless steel cup holds the sample. Another stainless steel electrode weight sits on top of the sample (direct contact with the dust layer). As voltage is increased from small amounts (e.g. 20 V), no current is measured. Then, a threshold voltage level is reached. At this level, current surges through the sample... so much so that the voltage supply unit can trip off. After removal of the unburned combustibles during the above-mentioned annealing procedure, the descending temperature mode curve shows the typical inverted "V" shape one might expect.

Low Resistivity

Particles that have low resistivity are difficult to collect because they are easily charged (very conductive) and rapidly lose their charge on arrival at the collection electrode. The particles take on the charge of the collection electrode, bounce off the plates, and become re-entrained in the gas stream. Thus, attractive and repulsive electrical forces that are normally at work at normal and higher resistivities are lacking, and the binding forces to the plate are considerably lessened. Examples of low-resistivity dusts are unburned carbon in fly ash and carbon black.

If these conductive particles are coarse, they can be removed upstream of the precipitator by using a device such as a cyclone mechanical collector.

The addition of liquid ammonia (NH_3) into the gas stream as a conditioning agent has found wide use in recent years. It is theorized that ammonia reacts with H_2SO_4 contained in the flue gas to form an ammonium sulfate compound that increases the cohesivity of the dust. This additional cohesivity makes up for the loss of electrical attraction forces.

The table below summarizes the characteristics associated with low, normal and high resistivity dusts.

The moisture content of the flue gas stream also affects particle resistivity. Increasing the moisture content of the gas stream by spraying water or injecting steam into the duct work preceding the ESP lowers the resistivity. In both temperature adjustment and moisture conditioning, one must maintain gas conditions above the dew point to prevent corrosion problems in the ESP or downstream equipment. The figure to the right shows the effect of temperature and moisture on the resistivity of a cement dust. As the percentage of moisture in the gas stream increases from 6 to 20%, the resistivity of the dust dramatically decreases. Also, raising or lowering the temperature can decrease cement dust resistivity for all the moisture percentages represented.

The presence of SO_3 in the gas stream has been shown to favor the electrostatic precipitation process when problems with high resistivity occur. Most of the sulfur content in the coal burned for combustion sources converts to SO_2. However, approximately 1% of the sulfur converts to SO_3. The amount of SO_3 in the flue gas normally increases with increasing sulfur content of the coal. The resistivity of the particles decreases as the sulfur content of the coal increases.

Resistivity	Range of Measurement	Precipitator Characteristics
Low	between 10^4 and 10^7 ohm-cm	1. Normal operating voltage and current levels unless dust layer is thick enough to reduce plate clearances and cause higher current levels. 2. Reduced electrical force component retaining collected dust, vulnerable to high reentrainment losses. 3. Negligible voltage drop across dust layer. 4. Reduced collection performance due to (2)
Normal	between 10^7 and 2×10^{10} ohm-cm	1. Normal operating voltage and current levels. 2. Negligible voltage drop across dust layer. 3. Sufficient electrical force component retaining collected dust. 4. High collection performance due to (1), (2) and (3)
Marginal to High	between 2×10^{10} and 10^{12} ohm-cm	1. Reduced operating voltage and current levels with high spark rates. 2. Significant voltage loss across dust layer. 3. Moderate electrical force component retaining collected dust. 4. Reduced collection performance due to (1) and (2)
High	above 10^{12} ohm-cm	1. Reduced operating voltage levels; high operating current levels if power supply controller is not operating properly. 2. Very significant voltage loss across dust layer. 3. High electrical force component retaining collected dust. 4. Seriously reduced collection performance due to (1), (2) and probably back corona.

Other conditioning agents, such as sulfuric acid, ammonia, sodium chloride, and soda ash (sometimes as raw trona), have also been used to reduce particle resistivity. Therefore, the chemical composition of the flue gas stream is important with regard to the resistivity of the particles to be collected in the ESP. The table below lists various conditioning agents and their mechanisms of operation.

Conditioning Agent	Mechanism(s) of Action
Sulfur Trioxide and/or Sulfuric Acid	1. Condensation and adsorption on fly ash surfaces. 2. may also increase cohesiveness of fly ash. 3. Reduces resistivity
Ammonia	Mechanism is not clear, various ones proposed; 1. Modifies resistivity. 2. Increases ash cohesiveness. 3. Enhances space charge effect.

Ammonium Sulfate	Little is known about the mechanism; claims are made for the following: 1. Modifies resistivity (depends upon injection temperature). 2. Increases ash cohesiveness. 3. Enhances space charge effect. 4. Experimental data lacking to substantiate which of these is predominant.
Triethylamine	Particle agglomeration claimed; no supporting data.
Sodium Compounds	1. Natural conditioner if added with coal. 2. Resistivity modifier if injected into gas stream.
Compounds of Transition Metals	Postulated that they catalyze oxidation of SO_2 to SO_3; no definitive tests with fly ash to verify this postulation.
Potassium Sulfate and Sodium Chloride	In cement and lime kiln ESPs: 1. Resistivity modifiers in the gas stream. 2. NaCl - natural conditioner when mixed with coal.

If injection of ammonium sulfate occurs at a temperature greater than about 600 °F (320 °C), dissociation into ammonia and sulfur trioxide results. Depending on the ash, SO_2 may preferentially interact with fly ash as SO_3 conditioning. The remainder recombines with ammonia to add to the space charge as well as increase cohesiveness of the ash.

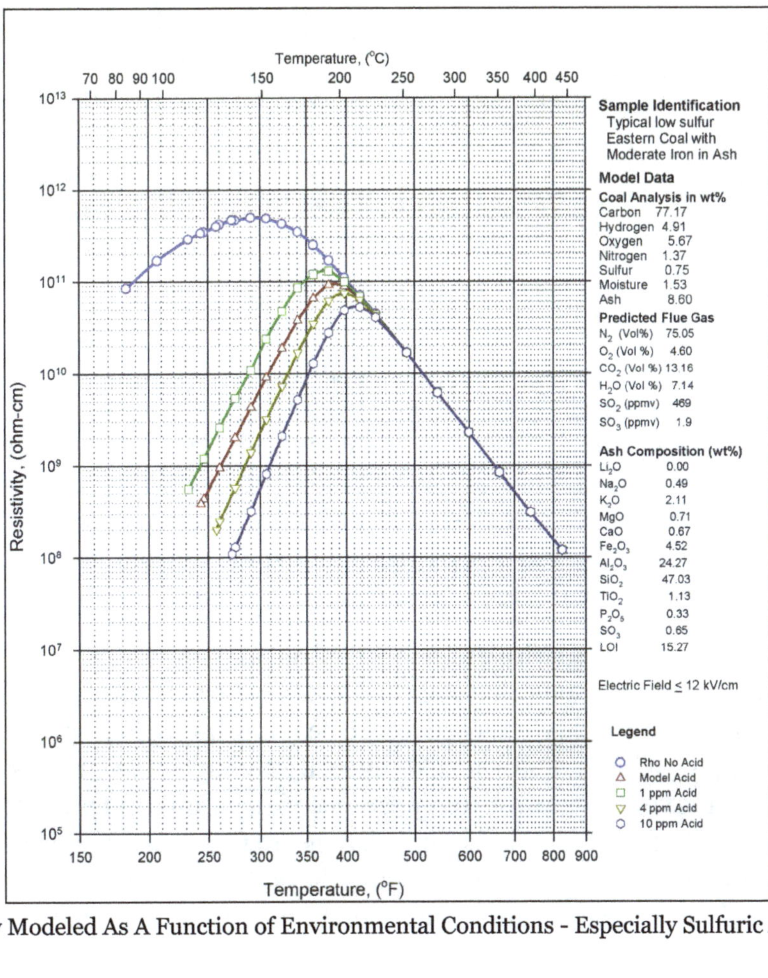

Resistivity Modeled As A Function of Environmental Conditions - Especially Sulfuric Acid Vapor.

More recently, it has been recognized that a major reason for loss of efficiency of the electrostatic precipitator is due to particle buildup on the charging wires in addition to the collection plates (Davidson and McKinney, 1998). This is easily remedied by making sure that the wires themselves are cleaned at the same time that the collecting plates are cleaned.

Sulfuric acid vapor (SO_3) enhances the effects of water vapor on surface conduction. It is physically adsorbed within the layer of moisture on the particle surfaces. The effects of relatively small amounts of acid vapor can be seen in the figure below and to the right.

The inherent resistivity of the sample at 300 °F (150 °C) is 5×10^{12} ohm-cm. An equilibrium concentration of just 1.9 ppm sulfuric acid vapor lowers that value to about 7×10^9 ohm-cm.

Modern Industrial Electrostatic Precipitators

A smokestack at coal-fired Hazelwood Power Station in Victoria, Australia emits brown smoke when its ESP is shut down.

ESPs continue to be excellent devices for control of many industrial particulate emissions, including smoke from electricity-generating utilities (coal and oil fired), salt cake collection from black liquor boilers in pulp mills, and catalyst collection from fluidized bed catalytic cracker units in oil refineries to name a few. These devices treat gas volumes from several hundred thousand ACFM to 2.5 million ACFM (1,180 m^3/s) in the largest coal-fired boiler applications. For a coal-fired boiler the collection is usually performed downstream of the air preheater at about 160 °C (320 °F) which provides optimal resistivity of the coal-ash particles. For some difficult applications with low-sulfur fuel hot-end units have been built operating above 370 °C (698 °F).

The original parallel plate–weighted wire design has evolved as more efficient (and robust) discharge electrode designs were developed, today focusing on rigid (pipe-frame) discharge electrodes to which many sharpened spikes are attached (barbed wire), maximizing corona production. Transformer-rectifier systems apply voltages of 50–100 kV at relatively high current densities. Modern controls, such as an automatic voltage control, minimize electric sparking and prevent arcing (sparks are quenched within 1/2 cycle of the TR set), avoiding damage to the components. Automatic plate-rapping systems and hopper-evacuation systems remove the collected particulate matter while on line, theoretically allowing ESPs to stay in continuous operation for years at a time.

Electrostatic Air Sampling

Electrostatic precipitators can be used to sample airborne particles or aerosol for analysis purposes. Precipitator designs with a liquid counterelectrode can be used to sample biological particles, e.g. viruses, directly into a small liquid volume to reduce unnecessary sample dilution.

Wet Electrostatic Precipitator

A wet electrostatic precipitator (WESP or wet ESP) operates with water vapor saturated air streams (100% relative humidity). WESPs are commonly used to remove liquid droplets such as sulfuric acid mist from industrial process gas streams. The WESP is also commonly used where the gases are high in moisture content, contain combustible particulate, or have particles that are sticky in nature.

The preferred and most modern type of WESP is a downflow tubular design. This design allows the collected moisture and particulate to form a moving slurry that helps to keep the collection surfaces clean. Plate style and upflow design WESPs are very unreliable and should not be used in applications where particulate is sticky in nature.

Consumer-oriented Electrostatic Air Cleaners

A portable electrostatic air cleaner marketed to consumers.

Plate precipitators are commonly marketed to the public as air purifier devices or as a permanent replacement for furnace filters, but all have the undesirable attribute of being somewhat messy to clean. A negative side-effect of electrostatic precipitation devices is the potential production of toxic ozone and NO_x. However, electrostatic precipitators offer benefits over other air purifications technologies, such as HEPA filtration, which require expensive filters and can become "production sinks" for many harmful forms of bacteria.

With electrostatic precipitators, if the collection plates are allowed to accumulate large amounts of particulate matter, the particles can sometimes bond so tightly to the metal plates that vigorous washing and scrubbing may be required to completely clean the collection plates. The close spacing of the plates can make thorough cleaning difficult, and the stack of plates often cannot be easily disassembled for cleaning. One solution, suggested by several manufacturers, is to wash the collector plates in a dishwasher.

Portable electrostatic air cleaner with cover removed, showing collector plates.

Some consumer precipitation filters are sold with special soak-off cleaners, where the entire plate array is removed from the precipitator and soaked in a large container overnight, to help loosen the tightly bonded particulates.

A study by the Canada Mortgage and Housing Corporation testing a variety of forced-air furnace filters found that ESP filters provided the best, and most cost-effective means of cleaning air using a forced-air system.

The first portable electrostatic air filter systems for homes was marketed in 1954 by Raytheon.

Dust Collector

A dust collector is a system used to enhance the quality of air released from industrial and commercial processes by collecting dust and other impurities from air or gas. Designed to handle high-volume dust loads, a dust collector system consists of a blower, dust filter, a filter-cleaning system, and a dust receptacle or dust removal system. It is distinguished from air cleaners, which use disposable filters to remove dust.

Baghouse Dust Collector for Asphalt Plants.

Two rooftop dust collectors in Pristina, Kosovo.

History

Wilhelm Beth.

„Beth"-Filter „KS" (1910).

The father of the dust collector was Wilhelm Beth from Lübeck.

Uses

Dust collectors are used in many processes to either recover valuable granular solid or powder from process streams, or to remove granular solid pollutants from exhaust gases prior to venting to the atmosphere. Dust collection is an online process for collecting any process-generated dust from the source point on a continuous basis. Dust collectors may be of single unit construction, or a collection of devices used to separate particulate matter from the process air. They are often used as an air pollution control device to maintain or improve air quality.

Mist collectors remove particulate matter in the form of fine liquid droplets from the air. They are often used for the collection of metal working fluids, and coolant or oil mists. Mist collectors are often used to improve or maintain the quality of air in the workplace environment.

Fume and smoke collectors are used to remove sub-micrometer-size particulates from the air. They effectively reduce or eliminate particulate matter and gas streams from many industrial processes such as welding, rubber and plastic processing, high speed machining with coolants, tempering, and quenching.

Types of Dust Collectors

Five main types of industrial dust collectors are:

- Inertial separators
- Fabric filters

- Wet scrubbers
- Unit collectors
- Electrostatic precipitators

Inertial Separators

Inertial separators separate dust from gas streams using a combination of forces, such as centrifugal, gravitational, and inertial. These forces move the dust to an area where the forces exerted by the gas stream are minimal. The separated dust is moved by gravity into a hopper, where it is temporarily stored.

The three primary types of inertial separators are:

- Settling chambers
- Baffle chambers
- Centrifugal collectors

Neither settling chambers nor baffle chambers are commonly used in the minerals processing industry. However, their principles of operation are often incorporated into the design of more efficient dust collectors.

Settling Chamber

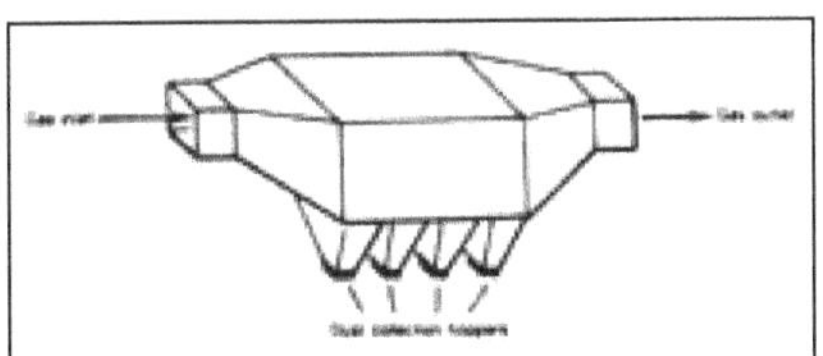

Settling Chamber

A settling chamber consists of a large box installed in the ductwork. The increase of cross section area at the chamber reduces the speed of the dust-filled airstream and heavier particles settle out.

Settling chambers are simple in design and can be manufactured from almost any material. However, they are seldom used as primary dust collectors because of their large space requirements and low efficiency. A practical use is as precleaners for more efficient collectors.

Baffle Chamber

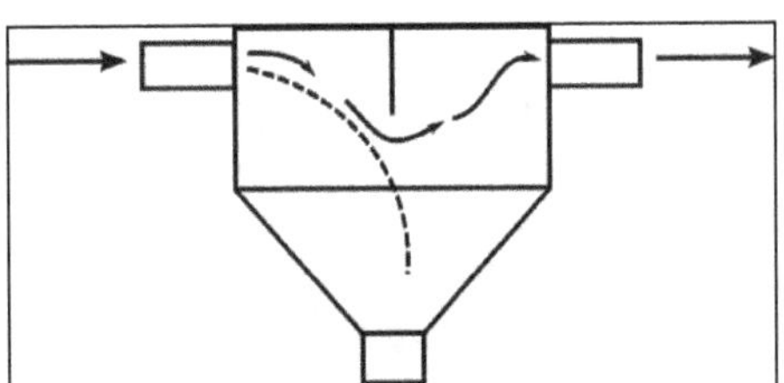

Baffle Chamber

Diagram of a Baffle Chamber.

Baffle chambers use a fixed baffle plate that causes the conveying gas stream to make a sudden change of direction. Large-diameter particles do not follow the gas stream but continue into a dead air space and settle. Baffle chambers are used as precleaners.

Centrifugal Collectors

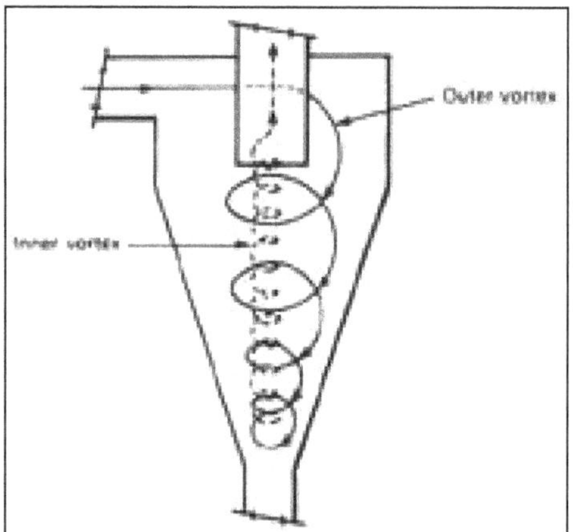

Cyclone

Centrifugal collectors use cyclonic action to separate dust particles from the gas stream. In a typical cyclone, the dust gas stream enters at an angle and is spun rapidly. The centrifugal force created by the circular flow throws the dust particles toward the wall of the cyclone. After striking the wall, these particles fall into a hopper located underneath.

The most common types of centrifugal, or inertial, collectors in use today are:-

Single-cyclone Separators

They create a dual vortex to separate coarse from fine dust. The main vortex spirals downward and carries most of the coarser dust particles. The inner vortex, created near the bottom of the cyclone, spirals upward and carries finer dust particles.

Multiple-cyclone Separators

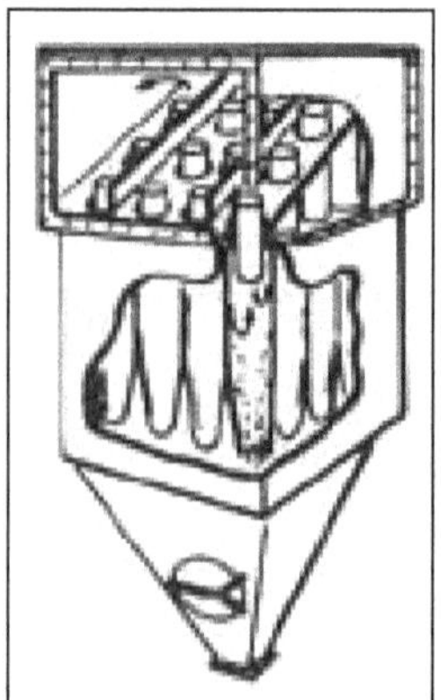

Multiclone

Multiple-cyclone separators consist of a number of small-diameter cyclones, operating in parallel and having a common gas inlet and outlet, as shown in the figure, and operate on the same principle as single cyclone separators—creating an outer downward vortex and an ascending inner vortex.

Multiple-cyclone separators remove more dust than single cyclone separators because the individual cyclones have a greater length and smaller diameter. The longer length provides longer residence time while the smaller diameter creates greater centrifugal force. These two factors result in better separation of dust particulates. The pressure drop of multiple-cyclone separators collectors is higher than that of single-cyclone separators, requiring more energy to clean the same amount of air. A single-chamber cyclone separator of the same volume is more economical, but doesn't remove as much dust.

Cyclone separators are found in all types of power and industrial applications, including pulp and paper plants, cement plants, steel mills, petroleum coke plants, metallurgical plants, saw mills and other kinds of facilities that process dust.

Secondary-air-flow Separators

This type of cyclone uses a secondary air flow, injected into the cyclone to accomplish several things. The secondary air flow increases the speed of the cyclonic action making the separator more efficient; it intercepts the particulate before it reaches the interior walls of the unit; and it forces the separated particulate toward the collection area. The secondary air flow protects the separator from particulate abrasion and allows the separator to be installed horizontally because gravity is not depended upon to move the separated particulate downward.

Fabric Filters

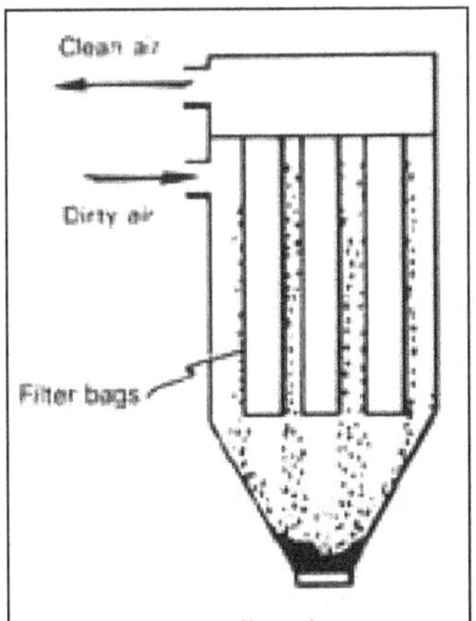

Baghouse

Commonly known as baghouses, fabric collectors use filtration to separate dust particulates from dusty gases. They are one of the most efficient and cost effective types of dust collectors available and can achieve a collection efficiency of more than 99% for very fine particulates.

Dust-laden gases enter the baghouse and pass through fabric bags that act as filters. The bags can be of woven or felted cotton, synthetic, or glass-fiber material in either a tube or envelope shape.

Pre-coating

To ensure the filter bags have a long usage life they are commonly coated with a filter enhancer (pre-coat). The use of chemically inert limestone (calcium carbonate) is most common as it maximises efficiency of dust collection (including fly ash) via formation of what is called a dustcake or coating on the surface of the filter media. This not only traps fine particulates but also provides protection for the bag itself from moisture, and oily or sticky particulates which can bind the filter media. Without a pre-coat the filter bag allows fine particulates to bleed through the bag filter system, especially during start-up, as the bag can only do part of the filtration leaving the finer parts to the filter enhancer dustcake.

Parts

Fabric filters generally have the following parts:

1. Clean plenum
2. Dusty plenum
3. Bag, cage, venturi assembly
4. Tubeplate
5. RAV/SCREW
6. Compressed air header
7. Blow pipe
8. Housing and hopper

Types of Bag Cleaning

Baghouses are characterized by their cleaning method.

Shaking

A rod connecting to the bag is powered by a motor. This provides motion to remove caked-on particles. The speed and motion of the shaking depends on the design of the bag and composition of the particulate matter. Generally shaking is horizontal. The top of the bag is closed and the bottom is open. When shaken, the dust collected on the inside of the bag is freed. During the cleaning process, no dirty gas flows through a bag while the bag is being cleaned. This redirection of air flow illustrates why baghouses must be compartmentalized.

Reverse Air

Air flow gives the bag structure. Dirty air flows through the bag from the inside, allowing dust to collect on the interior surface. During cleaning, gas flow is restricted from a specific compartment. Without the flowing air, the bags relax. The cylindrical bag contains rings that prevent it from completely collapsing under the pressure of the air. A fan blows clean air in the reverse direction. The relaxation and reverse air flow cause the dust cake to crumble and release into the hopper. Upon the completion of the cleaning process, dirty air flow continues and the bag regains its shape.

Pulse Jet

This type of baghouse cleaning (also known as pressure-jet cleaning) is the most common. A high pressure blast of air is used to remove dust from the bag. The blast enters the top of the bag tube, temporarily ceasing the flow of dirty air. The shock of air causes a wave of expansion to travel down the fabric. The flexing of the bag shatters and discharges the dust cake. The air burst is about 0.1 second and it takes about 0.5 seconds for the shock wave to travel down the length of the bag. Due to its rapid release, the blast of air does not interfere with contaminated gas flow. Therefore, pulse-jet baghouses can operate continuously and are not usually compartmentalized. The blast of compressed air must be powerful enough to ensure that the shock wave will travel the entire length of the bag and fracture the dust cake.

Sonic

The least common type of cleaning method is sonic. Shaking is achieved by sonic vibration. A sound generator produces a low frequency sound that causes the bags to vibrate. Sonic cleaning is commonly combined with another method of cleaning to ensure thorough cleaning.

Cartridge Collectors

Cartridge collectors use perforated metal cartridges that contain a pleated, nonwoven filtering media, as opposed to woven or felt bags used in baghouses. The pleated design allows for a greater total filtering surface area than in a conventional bag of the same diameter, The greater filtering area results in a reduced air to media ratio, pressure drop, and overall collector size.

Cartridge collectors are available in single use or continuous duty designs. In single-use collectors, the dirty cartridges are changed and collected dirt is removed while the collector is off. In the continuous duty design, the cartridges are cleaned by the conventional pulse-jet cleaning system.

Wet Scrubbers

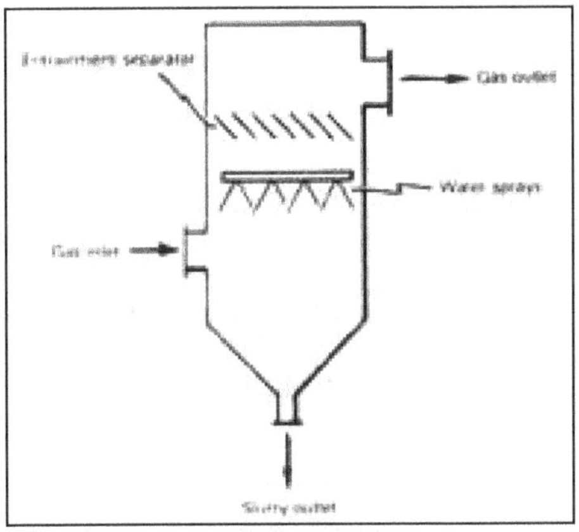

Dust collectors that use liquid are known as wet scrubbers. In these systems, the scrubbing liquid (usually water) comes into contact with a gas stream containing dust particles. Greater contact of the gas and liquid streams yields higher dust removal efficiency.

There is a large variety of wet scrubbers; however, all have one of three basic configurations:

1. Gas-humidification - The gas-humidification process agglomerates fine particles, increasing the bulk, making collection easier.

2. Gas-liquid contact - This is one of the most important factors affecting collection efficiency. The particle and droplet come into contact by four primary mechanisms:

a) Inertial impaction - When water droplets placed in the path of a dust-laden gas stream, the stream separates and flows around them. Due to inertia, the larger dust particles will continue on in a straight path, hit the droplets, and become encapsulated.

b) Interception - Finer particles moving within a gas stream do not hit droplets directly but brush against and adhere to them.

c) Diffusion - When liquid droplets are scattered among dust particles, the particles are deposited on the droplet surfaces by Brownian movement, or diffusion. This is the principal mechanism in the collection of submicrometre dust particles.

d) Condensation nucleation - If a gas passing through a scrubber is cooled below the dewpoint, condensation of moisture occurs on the dust particles. This increase in particle size makes collection easier.

3. Gas-liquid separation - Regardless of the contact mechanism used, as much liquid and dust as possible must be removed. Once contact is made, dust particulates and water droplets combine to form agglomerates. As the agglomerates grow larger, they settle into a collector.

The "cleaned" gases are normally passed through a mist eliminator (demister pads) to remove water droplets from the gas stream. The dirty water from the scrubber system is either cleaned and discharged or recycled to the scrubber. Dust is removed from the scrubber in a clarification unit or a drag chain tank. In both systems solid material settles on the bottom of the tank. A drag chain system removes the sludge and deposits in into a dumpster or stockpile.

Types of Scrubbers

Spray-tower scrubber wet scrubbers may be categorized by pressure drop as follows:

- Low-energy scrubbers (0.5 to 2.5 inches water gauge - 124.4 to 621.9 Pa)
- Low- to medium-energy scrubbers (2.5 to 6 inches water gauge - 0.622 to 1.493 kPa)
- Medium- to high-energy scrubbers (6 to 15 inches water gauge - 1.493 to 3.731 kPa)
- High-energy scrubbers (greater than 15 inches water gauge - greater than 3.731 kPa)

Due to the large number of commercial scrubbers available, it is not possible to describe each individual type here. However, the following sections provide examples of typical scrubbers in each category.

Low-energy Scrubbers

In the simple, gravity-spray-tower scrubber, liquid droplets formed by liquid atomized in spray nozzles fall through rising exhaust gases. Dirty water is drained at the bottom.

These scrubbers operated at pressure drops of 1 to 2 in. water gauge (¼ to ½ kPa) and are approximately 70% efficient on 10 μm particles. Their efficiency is poor below 10 μm. However, they are capable of treating relatively high dust concentrations without becoming plugged.

Low- to Medium-energy Scrubbers

Wet cyclones use centrifugal force to spin the dust particles (similar to a cyclone), and throw the particulates upon the collector's wetted walls. Water introduced from the top to wet the cyclone walls carries these particles away. The wetted walls also prevent dust reentrainment.

Pressure drops for these collectors range from 2 to 8 in. water (½ to 2 kPa), and the collection efficiency is good for 5 μm particles and above.

High-energy Scrubbers co-current-flow Scrubber

Packed-bed scrubbers consist of beds of packing elements, such as coke, broken rock, rings, saddles, or other manufactured elements. The packing breaks down the liquid flow into a high-surface-area film so that the dusty gas streams passing through the bed achieve maximum contact with the liquid film and become deposited on the surfaces of the packing elements. These scrubbers have a good collection efficiency for respirable dust.

Three types of packed-bed scrubbers are-

- Cross-flow scrubbers
- Co-current flow scrubbers
- Counter-current flow scrubbers

Efficiency can be greatly increased by minimizing target size, i.e., using 0.003 in. (0.076 mm) diameter stainless steel wire and increasing gas velocity to more than 1,800 ft/min (9.14 m/s).

High-energy Scrubbers

Venturi scrubbers consist of a venturi-shaped inlet and separator. The dust-laden gases venturi scrubber enter through the venturi and are accelerated to speeds between 12,000 and 36,000 ft/min (60.97-182.83 m/s). These high-gas velocities immediately atomize the coarse water spray, which is injected radially into the venturi throat, into fine droplets. High energy and extreme turbulence promote collision between water droplets and dust particulates in the throat. The agglomeration process between particle and droplet continues in the diverging section of the venturi. The large agglomerates formed in the venturi are then removed by an inertial separator.

Venturi scrubbers achieve very high collection efficiencies for respirable dust. Since efficiency of a venturi scrubber depends on pressure drop, some manufacturers supply a variable-throat venturi to maintain pressure drop with varying gas flows.

Electrostatic Precipitators (ESP)

Electrostatic precipitators use electrostatic forces to separate dust particles from exhaust gases. A number of high-voltage, direct-current discharge electrodes are placed between grounded collecting electrodes. The contaminated gases flow through the passage formed by the discharge and collecting electrodes. Electrostatic precipitators operate on the same principle as home "Ionic" air purifiers.

The airborne particles receive a negative charge as they pass through the ionized field between the electrodes. These charged particles are then attracted to a grounded or positively charged electrode and adhere to it.

The collected material on the electrodes is removed by rapping or vibrating the collecting electrodes either continuously or at a predetermined interval. Cleaning a precipitator can usually be done without interrupting the airflow.

The four main components of all electrostatic precipitators are:

- Power supply unit, to provide high-voltage DC power
- Ionizing section, to impart a charge to particulates in the gas stream
- A means of removing the collected particulates
- A housing to enclose the precipitator zone

The following factors affect the efficiency of electrostatic precipitators:

- Larger collection-surface areas and lower gas-flow rates increase efficiency because of the increased time available for electrical activity to treat the dust particles.
- An increase in the dust-particle migration velocity to the collecting electrodes increases efficiency. The migration velocity can be increased by:
    - Decreasing the gas viscosity
    - Increasing the gas temperature
    - Increasing the voltage field

Types of Precipitators

There are two main types of precipitators:

- High-voltage, single-stage - Single-stage precipitators combine an ionization and a collection step. They are commonly referred to as Cottrell precipitators.
- Low-voltage, two-stage - Two-stage precipitators use a similar principle; however, the ionizing section is followed by collection plates.

Described below is the high-voltage, single-stage precipitator, which is widely used in minerals processing operations. The low-voltage, two-stage precipitator is generally used for filtration in air-conditioning systems.

Plate Precipitators

The majority of electrostatic precipitators installed are the plate type. Particles are collected on flat, parallel surfaces that are 8 to 12 in. (20 to 30 cm) apart, with a series of discharge electrodes spaced along the centerline of two adjacent plates. The contaminated gases pass through the passage between the plates, and the particles become charged and adhere to the collection plates. Collected particles are usually removed by rapping the plates and deposited in bins or hoppers at the base of the precipitator.

Tubular Precipitators

Tubular precipitators consist of cylindrical collection electrodes with discharge electrodes located on the axis of the cylinder. The contaminated gases flow around the discharge electrode and up through the inside of the cylinders. The charged particles are collected on the grounded walls of the cylinder. The collected dust is removed from the bottom of the cylinder.

Tubular precipitators are often used for mist or fog collection or for adhesive, sticky, radioactive, or extremely toxic materials.

Unit Collectors

Unlike central collectors, unit collectors control contamination at its source. They are small and self-contained, consisting of a fan and some form of dust collector. They are suitable for isolated, portable, or frequently moved dust-producing operations, such as bins and silos or remote belt-conveyor transfer points. Advantages of unit collectors include small space requirements, the return of collected dust to main material flow, and low initial cost. However, their dust-holding and storage capacities, servicing facilities, and maintenance periods have been sacrificed.

A number of designs are available, with capacities ranging from 200 to 2,000 ft^3/min (90 to 900 L/s). There are two main types of unit collectors:

- Fabric collectors, with manual shaking or pulse-jet cleaning - normally used for fine dust
- Cyclone collectors - normally used for coarse dust

Fabric collectors are frequently used in minerals processing operations because they provide high collection efficiency and uninterrupted exhaust airflow between cleaning cycles. Cyclone collectors are used when coarser dust is generated, as in woodworking, metal grinding, or machining.

The following points should be considered when selecting a unit collector:

- Cleaning efficiency must comply with all applicable regulations.
- The unit maintains its rated capacity while accumulating large amounts of dust between cleanings.

- Simple cleaning operations do not increase the surrounding dust concentration.
- Has the ability to operate unattended for extended periods of time (for example, 8 hours).
- Automatic discharge or sufficient dust storage space to hold at least one week's accumulation.
- If renewable filters are used, they should not have to be replaced more than once a month.
- Durable.
- Quiet.

Use of unit collectors may not be appropriate if the dust-producing operations are located in an area where central exhaust systems would be practical. Dust removal and servicing requirements are expensive for many unit collectors and are more likely to be neglected than those for a single, large collector.

Selecting a Dust Collector

Dust collectors vary widely in design, operation, effectiveness, space requirements, construction, and capital, operating, and maintenance costs. Each type has advantages and disadvantages. However, the selection of a dust collector should be based on the following general factors:

- Dust concentration and particle size - For minerals processing operations, the dust concentration can range from 0.1 to 5.0 grains (0.32 g) of dust per cubic feet of air (0.23 to 11.44 grams per standard cubic meter), and the particle size can vary from 0.5 to 100 micrometres (μm) in diameter.
- Degree of dust collection required - The degree of dust collection required depends on its potential as a health hazard or public nuisance, the plant location, the allowable emission rate, the nature of the dust, its salvage value, and so forth. The selection of a collector should be based on the efficiency required and should consider the need for high-efficiency, high-cost equipment, such as electrostatic precipitators; high-efficiency, moderate-cost equipment, such as baghouses or wet scrubbers; or lower cost, primary units, such as dry centrifugal collectors.
- Characteristics of airstream - The characteristics of the airstream can have a significant impact on collector selection. For example, cotton fabric filters cannot be used where air temperatures exceed 180 °F (82 °C). Also, condensation of steam or water vapor can blind bags. Various chemicals can attack fabric or metal and cause corrosion in wet scrubbers.
- Characteristics of dust - Moderate to heavy concentrations of many dusts (such as dust from silica sand or metal ores) can be abrasive to dry centrifugal collectors. Hygroscopic material can blind bag collectors. Sticky material can adhere to collector elements and plug passages. Some particle sizes and shapes may rule out certain types of fabric collectors. The combustible nature of many fine materials rules out the use of electrostatic precipitators.
- Methods of disposal - Methods of dust removal and disposal vary with the material, plant process, volume, and type of collector used. Collectors can unload continuously or in batches. Dry

materials can create secondary dust problems during unloading and disposal that do not occur with wet collectors. Disposal of wet slurry or sludge can be an additional material-handling problem; sewer or water pollution problems can result if wastewater is not treated properly.

Fan and Motor

The fan and motor system supplies mechanical energy to move contaminated air from the dust-producing source to a dust collector.

Types of Fans

There are two main kinds of industrial fans:

- Centrifugal fans
- Axial-flow fans

Centrifugal Fans

Centrifugal fans consist of a wheel or a rotor mounted on a shaft that rotates in a scroll-shaped housing. Air enters at the eye of the rotor, makes a right-angle turn, and is forced through the blades of the rotor by centrifugal force into the scroll-shaped housing. The centrifugal force imparts static pressure to the air. The diverging shape of the scroll also converts a portion of the velocity pressure into static pressure.

There are three main types of centrifugal fans:

- Radial-blade fans - Radial-blade fans are used for heavy dust loads. Their straight, radial blades do not get clogged with material, and they withstand considerable abrasion. These fans have medium tip speeds and medium noise factors.
- Backward-blade fans - Backward-blade fans operate at higher tip speeds and thus are more efficient. Since material may build up on the blades, these fans should be used after a dust collector. Although they are noisier than radial-blade fans, backward-blade fans are commonly used for large-volume dust collection systems because of their higher efficiency.
- Forward-curved-blade fans - These fans have curved blades that are tipped in the direction of rotation. They have low space requirements, low tip speeds, and a low noise factor. They are usually used against low to moderate static pressures.

Axial-flow Fans

Axial-flow fans are used in systems that have low resistance levels. These fans move the air parallel to the fan's axis of rotation. The screw-like action of the propellers moves the air in a straight-through parallel path, causing a helical flow pattern.

The three main kinds of axial fans are-

- Propeller fans - These fans are used to move large quantities of air against very low static pressures. They are usually used for general ventilation or dilution ventilation and are good in developing up to 0.5 in. wg (124.4 Pa).

- Tube-axial fans - Tube-axial fans are similar to propeller fans except they are mounted in a tube or cylinder. Therefore, they are more efficient than propeller fans and can develop up to 3 to 4 in. wg (743.3 to 995 Pa). They are best suited for moving air containing substances such as condensible fumes or pigments.
- Vane-axial fans - Vane-axial fans are similar to tube-axial fans except air-straightening vanes are installed on the suction or discharge side of the rotor. They are easily adapted to multistaging and can develop static pressures as high as 14 to 16 in. wg (3.483 to 3.98 kPa). They are normally used for clean air only.

Fan Selection

When selecting a fan, the following points should be considered:

- Volume required
- Fan static pressure
- Type of material to be handled through the fan (For example, a radial-blade fan should be used with fibrous material or heavy dust loads, and nonsparking construction must be used with explosive or inflammable materials.)
- Type of drive arrangement, such as direct drive or belt drive
- Space requirements
- Noise levels
- Operating temperature (For example, sleeve bearings are suitable to 250 °F/121.1 °C; ball bearings to 550 °F/287.8 °C)
- Sufficient size to handle the required volume and pressure with minimum horsepower
- Need for special coatings or construction when operating in corrosive atmospheres
- Ability of fan to accommodate small changes in total pressure while maintaining the necessary air volume
- Need for an outlet damper to control airflow during cold starts (If necessary, the damper may be interlocked with the fan for a gradual start until steady-state conditions are reached.)

Fan Rating Tables

After the above information is collected, the actual selection of fan size and speed is usually made from a rating table published by the fan manufacturer. This table is known as a multirating table, and it shows the complete range of capacities for a particular size of fan.

Points to note:

- The multirating table shows the range of pressures and speeds possible within the limits of the fan's construction.
- A particular fan may be available in different construction classes (identified as class I through IV) relating to its capabilities and limits.
- For a given pressure, the highest mechanical efficiency is usually found in the middle third of the volume column.
- A fan operating at a given speed can have an infinite number of ratings (pressure and volume) along the length of its characteristic curve. However, when the fan is installed in a dust collection system, the point of rating can only be at the point at which the system resistance curve intersects the fan characteristic curve.
- In a given system, a fan at a fixed speed or at a fixed blade setting can have a single rating only. This rating can be changed only be changing the fan speed, blade setting, or the system resistance.
- For a given system, an increase in exhaust volume will result in increases in static and total pressures. For example, for a 20% increase in exhaust volume in a system with 5 in. pressure loss, the new pressure loss will be $5 \times (1.20)^2 = 7.2$ in.
- For rapid estimates of probable exhaust volumes available for a given motor size, the equation for brake horsepower, as illustrated, can be useful.

Fan installation Typical fan discharge conditions Fan ratings for volume and static pressure, as described in the multirating tables, are based on the tests conducted under ideal conditions. Often, field installation creates airflow problems that reduce the fan's air delivery. The following points should be considered when installing the fan:

- Avoid installation of elbows or bends at the fan discharge, which will lower fan performance by increasing the system's resistance.
- Avoid installing fittings that may cause non-uniform flow, such as an elbow, mitred elbow, or square duct.
- Check that the fan impeller is rotating in the proper direction-clockwise or counterclockwise.
- For belt-driven fans-
    - Check that the motor sheave and fan sheave are aligned properly.
    - Check for proper belt tension.
- Check the passages between inlets, impeller blades, and inside of housing for buildup of dirt, obstructions, or trapped foreign matter.

Electric Motors

Electric motors are used to supply the necessary energy to drive the fan.

Integral-horsepower electric motors are normally three-phase, alternating-current motors. Fractional-horsepower electric motors are normally single-phase, alternating-current motors and are used when less than 1 hp (0.75 kW) is required. Since most dust collection systems require motors with more than 1 hp (0.75 kW), only integral-horsepower motors are discussed here.

The two most common types of integral-horsepower motors used in dust collection systems are-

- Squirrel-cage motors - These motors have a constant speed and are of a nonsynchronous, induction type.
- Wound-rotor motors - These motors are also known as slip-ring motors. They are general-purpose or continuous-rated motors and are chiefly used when an adjustable-speed motor is desired.

Squirrel-cage and wound-rotor motors are further classified according to the type of enclosure they use to protect their interior windings. These enclosures fall into two broad categories:

- Open
- Totally enclosed

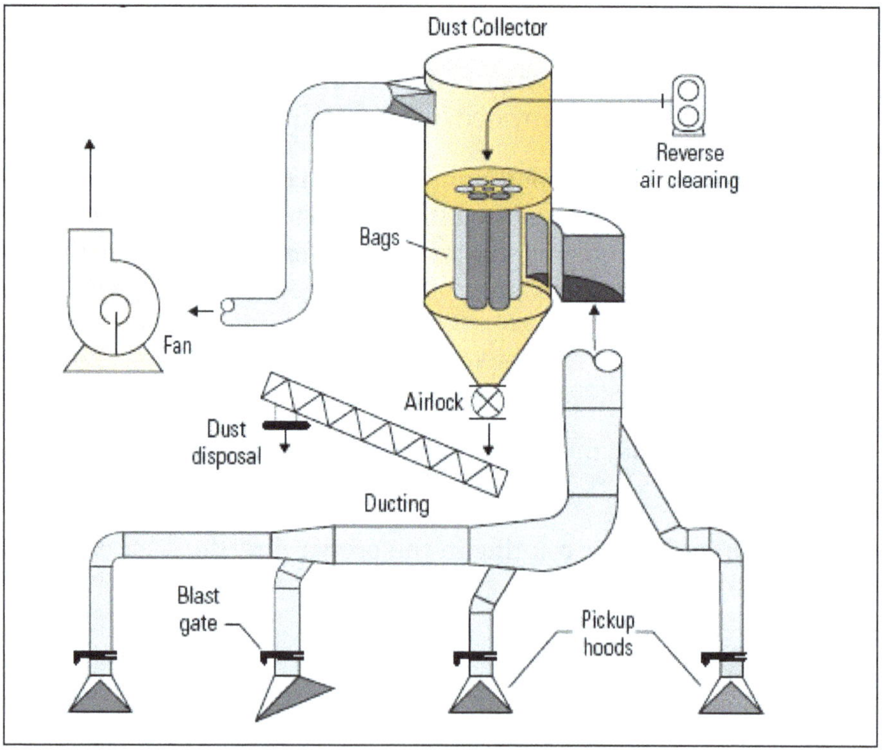

Figure 1. Dust Collection System Example.

Drip-proof and splash-proof motors are open motors. They provide varying degrees of protection; however, they should not be used where the air contains substances that might be harmful to the interior of the motor.

Totally enclosed motors are weather-protected with the windings enclosed. These enclosures prevent free exchange of air between the inside and the outside, but they are not airtight.

Totally enclosed, fan-cooled (TEFC) motors are another kind of totally enclosed motor. These motors are the most commonly used motors in dust collection systems. They have an integral-cooling fan outside the enclosure, but within the protective shield, that directs air over the enclosure.

Both open and totally enclosed motors are available in explosion-proof and dust-ignition-proof models to protect against explosion and fire in hazardous environments.

Motors are selected to provide sufficient power to operate fans over the full range of process conditions (temperature and flow rate).

Configurations

Dust collectors can be configured into one of five common types.

1. Ambient units - Ambient units are free-hanging systems for use when applications limit the use of source-capture arms or ductwork.
2. Collection booths - Collector booths require no ductwork, and allow the worker greater freedom of movement. They are often portable.
3. Downdraft tables - A downdraft table is a self-contained portable filtration system that removes harmful particulates and returns filtered air back into the facility with no external ventilation required.
4. Source collector or Portable units - Portable units are for collecting dust, mist, fumes, or smoke at the source.
5. Stationary units - An example of a stationary collector is a baghouse.

Parameters Involved in Specifying Dust Collectors

Important parameters in specifying dust collectors include airflow the velocity of the air stream created by the vacuum producer; system power, the power of the system motor, usually specified in horsepower; storage capacity for dust and particles, and minimum particle size filtered by the unit. Other considerations when choosing a dust collection system include the temperature, moisture content, and the possibility of combustion of the dust being collected.

Systems for fine removal may only contain a single filtration system (such as a filter bag or cartridge). However, most units utilize a primary and secondary separation/filtration system. In many cases the heat or moisture content of dust can negatively affect the filter media of a baghouse or cartridge dust collector. A cyclone separator or dryer may be placed before these units to reduce heat or moisture content before reaching the filters. Furthermore, some units may have third and fourth stage filtration. All separation and filtration systems used within the unit should be specified.

A baghouse is an air pollution abatement device used to trap particulate by filtering gas streams through large fabric bags. They are typically made of glass fibers or fabric.

An impinger system is a device in which particles are removed by impacting the aerosol particles into a liquid. Modular media type units combine a variety of specific filter modules in one unit. These systems can provide solutions to many air contaminant problems. A typical system incorporates a series of disposable or cleanable pre-filters, a disposable vee-bag or cartridge filter. HEPA or carbon final filter modules can also be added. Various models are available, including free-hanging or ducted installations, vertical or horizontal mounting, and fixed or portable configurations. Filter cartridges are made out of a variety of synthetic fibers and are capable of collecting sub-micrometre particles without creating an excessive pressure drop in the system. Filter cartridges require periodic cleaning.

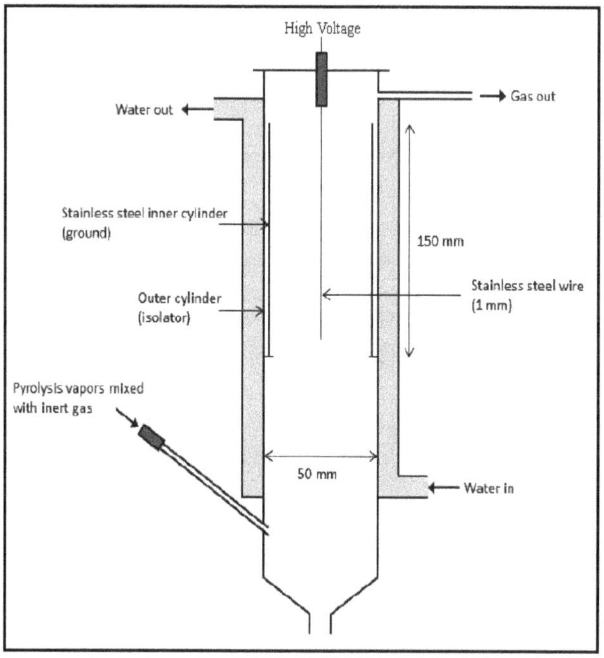

Electrostatic precipitators are a type of air cleaner, which charges particles of dust by passing dust-laden air through a strong (50-100 kV) electrostatic field. This causes the particles to be attracted to oppositely charged plates so that they can be removed from the air stream.

A wet scrubber, or venturi scrubber, is similar to a cyclone but it has an orifice unit that sprays water into the vortex in the cyclone section, collecting all of the dust in a slurry system. The water media can be recirculated and reused to continue to filter the air. Eventually the solids must be removed from the water stream and disposed of.

Filter Cleaning Methods

Online cleaning – automatically timed filter cleaning which allows for continuous, uninterrupted dust collector operation for heavy dust operations.

Offline cleaning – filter cleaning accomplished during dust collector shut down. Practical whenever the dust loading in each dust collector cycle does not exceed the filter capacity. Allows for maximum effectiveness in dislodging and disposing of dust.

On-demand cleaning – filter cleaning initiated automatically when the filter is fully loaded, as determined by a specified drop in pressure across the media surface.

Reverse-pulse/Reverse-jet cleaning – Filter cleaning method which delivers blasts of compressed air from the clean side of the filter to dislodge the accumulated dust cake.

Impact/Rapper cleaning – Filter cleaning method in which high-velocity compressed air forced through a flexible tube results in a random rapping of the filter to dislodge the dust cake. Especially effective when the dust is extremely fine or sticky.

Wet Scrubber

The term wet scrubber describes a variety of devices that remove pollutants from a furnace flue gas or from other gas streams. In a wet scrubber, the polluted gas stream is brought into contact with the scrubbing liquid, by spraying it with the liquid, by forcing it through a pool of liquid, or by some other contact method, so as to remove the pollutants.

Design

The design of wet scrubbers or any air pollution control device depends on the industrial process conditions and the nature of the air pollutants involved. Inlet gas characteristics and dust properties (if particles are present) are of primary importance. Scrubbers can be designed to collect particulate matter and/or gaseous pollutants. The versatility of wet scrubbers allow them to be built in numerous configurations, all designed to provide good contact between the liquid and polluted gas stream.

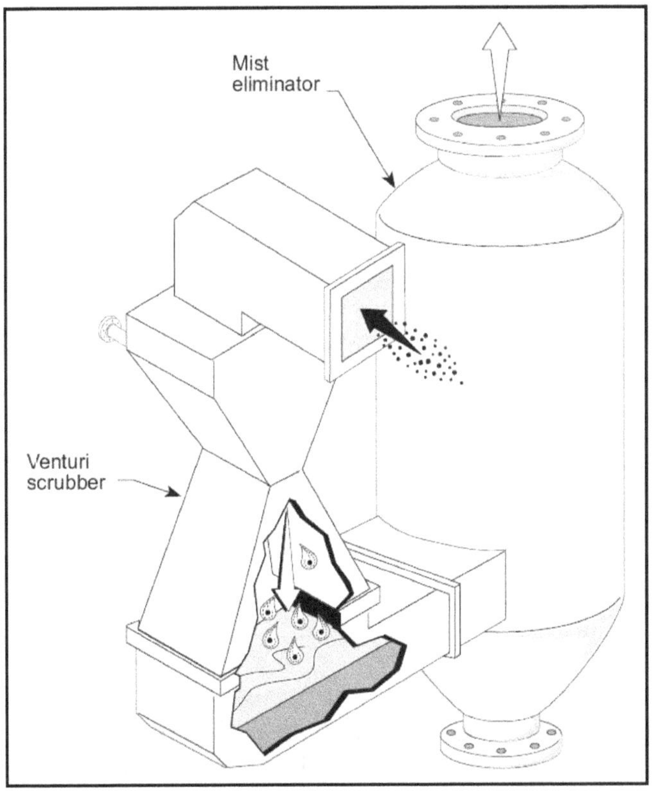

A venturi scrubber design. The mist eliminator for a venturi scrubber is often a separate device called a cyclonic separator.

Wet scrubbers remove dust particles by *capturing* them in liquid droplets. The droplets are then collected, the liquid *dissolving* or *absorbing* the pollutant gases. Any droplets that are in the scrubber inlet gas must be separated from the outlet gas stream by means of another device referred to as a mist eliminator or entrainment separator (these terms are interchangeable). Also, the resultant scrubbing liquid must be treated prior to any ultimate discharge or being reused in the plant.

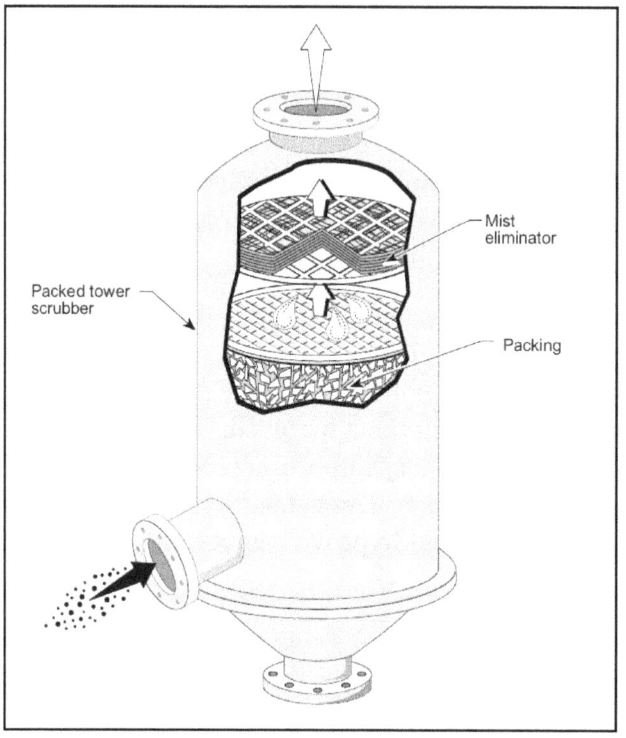

A packed bed tower design where the mist eliminator is built into the top of the structure. Various tower designs exist.

A wet scrubber's ability to collect small particles is often directly proportional to the power input into the scrubber. Low energy devices such as spray towers are used to collect particles larger than 5 micrometers. To obtain high efficiency removal of 1 micrometer (or less) particles generally requires high energy devices such as venturi scrubbers or augmented devices such as condensation scrubbers. Additionally, a properly designed and operated entrainment separator or mist eliminator is important to achieve high removal efficiencies. The greater the number of liquid droplets that are not captured by the mist eliminator, the higher the potential emission levels.

Wet scrubbers that remove gaseous pollutants are referred to as *absorbers*. Good gas-to-liquid contact is essential to obtain high removal efficiencies in absorbers. A number of wet scrubber designs are used to remove gaseous pollutants, with the packed tower and the plate tower being the most common.

If the gas stream contains both particle matter and gases, wet scrubbers are generally the only single air pollution control device that can remove both pollutants. Wet scrubbers can achieve high removal efficiencies for either particles or gases and, in some instances, can achieve a high removal efficiency for both pollutants in the same system. However, in many cases, the best operating conditions for particles collection are the poorest for gas removal.

In general, obtaining high simultaneous gas and particulate removal efficiencies requires that one of them be easily collected (i.e., that the gases are very soluble in the liquid or that the particles are large and readily captured) or by the use of a scrubbing reagent such as lime or sodium hydroxide.

Advantages and Disadvantages

For particulate control, wet scrubbers (also referred to as wet collectors) are evaluated against fabric filters and electrostatic precipitators (ESPs). Some *advantages* of wet scrubbers over these devices are as follows:

- Wet scrubbers have the ability to handle high temperatures and moisture.
- In wet scrubbers, flue gases are cooled, resulting in smaller overall size of equipment.
- Wet scrubbers can remove both gases and particulate matter.
- Wet scrubbers can neutralize corrosive gases.

Some *disadvantages* of wet scrubbers include corrosion, the need for entrainment separation or mist removal to obtain high efficiencies and the need for treatment or reuse of spent liquid.

Wet scrubbers have been used in a variety of industries such as acid plants, fertilizer plants, steel mills, asphalt plants, and large power plants.

Relative advantages and disadvantages of wet scrubbers compared to other control devices	
Advantages	**Disadvantages**
• Small space requirements: Scrubbers reduce the temperature and volume of the unsaturated exhaust stream. Therefore, vessel sizes, including fans and ducts downstream, are smaller than those of other control devices. Smaller sizes result in lower capital costs and more flexibility in site location of the scrubber.	• Corrosion problems: Water and dissolved pollutants can form highly corrosive acid solutions. Proper construction materials are very important. Also, wet-dry interface areas can result in corrosion.
• No secondary dust sources: Once particulate matter is collected, it cannot escape from hoppers or during transport.	• High power requirements: High collection efficiencies for particulate matter are attainable only at high pressure drops, resulting in high operating costs.
• Handles high-temperature, high-humidity gas streams: No temperature limits or condensation problems can occur as in baghouses or ESPs.	• Water-disposal problems: Settling ponds or sludge clarifiers may be needed to meet waste-water regulations.
• Minimal fire and explosion hazards: Various dry dusts are flammable. Using water eliminates the possibility of explosions.	• Difficult product recovery: Dewatering and drying of scrubber sludge make recovery of any dust for reuse very expensive and difficult.
• Ability to collect both gases and particulate matter.	

Components

Wet scrubber systems generally consist of the following components:

- Ductwork and fan system.
- A saturation chamber (optional).
- Scrubbing vessel.
- Entrainment separator or mist eliminator.

- Pumping (and possible recycle system).
- Spent scrubbing liquid treatment and/or reuse system.
- An exhaust stack.

A typical wet scrubbing process can be described as follows:

- Hot flue gas from a furnace enters a saturator (if present) where gases are cooled and humidified prior to entering the scrubbing area. The saturator removes a small percentage of the particulate matter present in the flue gas.
- Next, the gas enters a venturi scrubber where approximately half of the gases are removed. Venturi scrubbers have a minimum particle removal efficiency of 95%.
- The gas flows through a second scrubber, a packed bed absorber, where the rest of the gases (and particulate matter) are collected.
- An entrainment separator or mist eliminator removes any liquid droplets that may have become entrained in the flue gas.
- A recirculation pump moves some of the spent scrubbing liquid back to the venturi scrubber where it is recycled and the remainder is sent to a treatment system.
- Treated scrubbing liquid is recycled back to the saturator and the packed bed absorber.
- Fans and ductwork move the flue gas stream through the system and eventually out the stack.

Categorization

Since wet scrubbers vary greatly in complexity and method of operation, devising categories into which all of them neatly fit is extremely difficult. Scrubbers for particle collection are usually categorized by the gas-side pressure drop of the system. Gas-side pressure drop refers to the pressure difference, or pressure drop, that occurs as the exhaust gas is pushed or pulled through the scrubber, disregarding the pressure that would be used for pumping or spraying the liquid into the scrubber.

Scrubbers may be classified *by pressure drop* as follows:

- *Low-energy scrubbers* have pressure drops of less than 12.7 cm (5 in) of water.
- *Medium-energy scrubbers* have pressure drops between 12.7 and 38.1 cm (5 and 15 in) of water.
- *High-energy scrubbers* have pressure drops greater than 37.1 cm (15 in) of water.

However, most scrubbers operate over a wide range of pressure drops, depending on their specific application, thereby making this type of categorization difficult.

Another way to classify wet scrubbers is by their *use* - to primarily collect either particulates or gaseous pollutants. Again, this distinction is not always clear since scrubbers can often be used to remove both types of pollutants.

Wet scrubbers can also be categorized by the manner in which the gas and liquid phases are brought into contact. Scrubbers are designed to use power, or energy, from the gas stream or the liquid stream, or some other method to bring the pollutant gas stream into contact with the liquid. These categories are given in Table 2.

Categories of wet collectors by energy source used for contact	
Wet collector	**Energy source used for gas-liquid contact**
• Gas-phase contacting	• Gas stream
• Liquid-phase contacting	• Liquid stream
• Wet film	• Liquid and gas streams
• Combination	• Energy source:
○ Liquid phase and gas phase	○ Liquid and gas streams
○ Mechanically aided	○ Mechanically driven rotor

Material of Construction and Design

Corrosion can be a prime problem associated with chemical industry scrubbing systems. Fibre-reinforced plastic and dual laminates are often used as most dependable materials of construction.

References

- "Catalytic Oxidizer". U.S. EPA Technology Transfer Network Clearinghouse for Inventories & Emissions Factors. U.S. Environmental Protection Agency. Retrieved 11, February 2020
- Robert W., Hahn (November 2011). "The effect of Allowance Allocations on Cap-and-Trade System Performance" (PDF). Journal of Law and Economics. 54 (4). Retrieved 15, April 2020
- Burney, Nelson E. (2010). Carbon Tax and Cap-and-trade Tools : Market-based Approaches for Controlling Greenhouse Gases. New York: Nova Science Publishers, Inc. ISBN 9781608761371
- "Thermal Oxidizer". U.S. EPA Technology Transfer Network Clearinghouse for Inventories & Emissions Factors. U.S. Environmental Protection Agency. Retrieved 25, July 2020
- Dryzek, John S.; Norgaard, Richard B.; Schlosberg, David (2011). The Oxford Handbook of Climate Change and Society. Oxford University Press. p. 154. ISBN 9780199683420
- Voss, Jan-Peter (2007). "Innovation processes in governance: the development of emissions trading as a new policy instrument". Science and Public Policy. 34: 329–343. doi:10.3152/030234207x228584
- Tiwari, Gopal Nath; Agrawal, Basant (2010). Building integrated photovoltaic thermal systems : for sustainable developments. Cambridge: Royal Society of Chemistry. ISBN 1-84973-090-3
- Nikulshina, V.; Ayesa, N.; Gálvez, M. E.; Stainfeld, A. (2016). "Feasibility of Na–Based Thermochemical Cycles for the Capture of CO_2 from air. Thermodynamic and Thermogravimetric Analyses". Chem. Eng. J. 140 (1–3): 62–70. doi:10.1016/j.cej.2007.09.007

- "SWANA 2012 Excellence Award Application "Landfill Gas Control" Seneca Landfill, Inc" (PDF). Retrieved 25, June 2020
- Sinden, Amy (2014). "Cost-Benefit Analysis, Ben Franklin, and the Supreme Court" (PDF). UC Irvine Law Review. University of California, Irvine School of Law. Retrieved 04, August 2020
- Carlson, Curtis; Burtraw, Dallas; Cropper, Maureen; Palmer, Karen L. (2000). "Sulfur dioxide control by electric utilities: What are the gains from trade?". Journal of Political Economy. 108: 1292–1326. doi:10.1086/317681
- Vincent D. Blondel: Recent Advances in Learning and Control, p. 233, Springer Science & Business Media, 2008, ISBN 9781848001541

5

Ozone Depletion

Highly reactive gas composed of three oxygen atoms is called Ozone. It is of pale blue color and has a pungent smell. Ozone depletion can be termed as breaking of ozone layer which covers the topmost layer of Earth's atmosphere. This chapter examines ozone cycle, ozone layer depletion, CFCs, ozone hole and its consequences in a detailed manner.

Ozone depletion describes two distinct but related phenomena observed since the late 1970s: a steady decline of about four percent in the total amount of ozone in Earth's stratosphere (the ozone layer), and a much larger springtime decrease in stratospheric ozone around Earth's polar regions. The latter phenomenon is referred to as the ozone hole. In addition to these well-known stratospheric phenomena, there are also springtime polar tropospheric ozone depletion events.

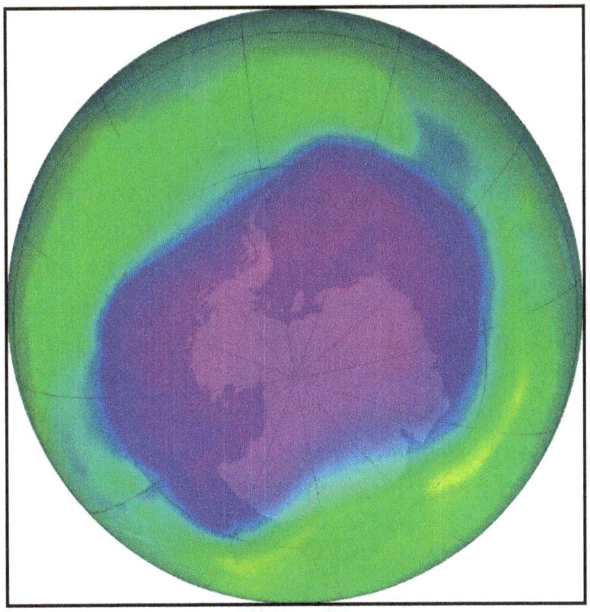

Image of the largest Antarctic ozone hole ever recorded (September 2006), over the Southern pole.

The details of polar ozone hole formation differ from that of mid-latitude thinning but the most important process in both is catalytic destruction of ozone by atomic halogens. The main source of these halogen atoms in the stratosphere is photodissociation of man-made halocarbon refrigerants, solvents, propellants, and foam-blowing agents (chlorofluorocarbon (CFCs), HCFCs, freons, halons). These compounds are transported into the stratosphere by winds after being emitted at the surface. Both types of ozone depletion were observed to increase as emissions of halocarbons increased.

CFCs and other contributory substances are referred to as ozone-depleting substances (ODS). Since the ozone layer prevents most harmful UVB wavelengths (280–315 nm) of ultraviolet light (UV light) from passing through the Earth's atmosphere, observed and projected decreases in ozone generated worldwide concern, leading to adoption of the Montreal Protocol that bans the production of CFCs, halons, and other ozone-depleting chemicals such as carbon tetrachloride and trichloroethane. It is suspected that a variety of biological consequences such as increases in sunburn, skin cancer, cataracts, damage to plants, and reduction of plankton populations in the ocean's photic zone may result from the increased UV exposure due to ozone depletion.

Ozone Cycle Overview

Three forms (or allotropes) of oxygen are involved in the ozone-oxygen cycle: oxygen atoms (O or atomic oxygen), oxygen gas (O_2 or diatomic oxygen), and ozone gas (O_3 or triatomic oxygen). Ozone is formed in the stratosphere when oxygen molecules photodissociate after intaking an ultraviolet photon whose wavelength is shorter than 240 nm. This converts a single O_2 into two atomic oxygen radicals. The atomic oxygen radicals then combine with separate O_2 molecules to create two O_3 molecules. These ozone molecules absorb ultraviolet (UV) light between 310 and 200 nm, following which ozone splits into a molecule of O_2 and an oxygen atom. The oxygen atom then joins up with an oxygen molecule to regenerate ozone. This is a continuing process that terminates when an oxygen atom "recombines" with an ozone molecule to make two O_2 molecules.

$2\ O_3 \rightarrow 3\ O_2$

The overall amount of ozone in the stratosphere is determined by a balance between photochemical production and recombination.

Ozone can be destroyed by a number of free radical catalysts, the most important of which are the hydroxyl radical (OH·), nitric oxide radical (NO·), chlorine radical (Cl·) and bromine radical (Br·). The dot is a common notation to indicate that all of these species have an unpaired electron and are thus extremely reactive. All of these have both natural and man-made sources; at the present time, most of the OH· and NO· in the stratosphere is of natural origin, but human activity has dramatically increased the levels of chlorine and bromine. These elements are found in certain stable organic compounds, especially chlorofluorocarbons, which may find their way to the stratosphere without being destroyed in the troposphere due to their low reactivity. Once in the stratosphere, the Cl and Br atoms are liberated from the parent compounds by the action of ultraviolet light, e.g.

$CFCl_3$ + electromagnetic radiation $\rightarrow$ Cl· + ·$CFCl_2$

The Cl and Br atoms can then destroy ozone molecules through a variety of catalytic cycles. In the simplest example of such a cycle, a chlorine atom reacts with an ozone molecule, taking an oxygen atom with it (forming ClO) and leaving a normal oxygen molecule. The chlorine monoxide (i.e., the ClO) can react with a second molecule of ozone (i.e., O_3) to yield another chlorine atom and two molecules of oxygen. The chemical shorthand for these gas-phase reactions is:

- Cl· + $O_3 \rightarrow$ ClO + O_2: The chlorine atom changes an ozone molecule to ordinary oxygen
- ClO + $O_3 \rightarrow$ Cl· + 2 O_2: The ClO from the previous reaction destroys a second ozone molecule and recreates the original chlorine atom, which can repeat the first reaction and continue to destroy ozone.

The overall effect is a decrease in the amount of ozone, though the rate of these processes can be decreased by the effects of null cycles. More complicated mechanisms have been discovered that lead to ozone destruction in the lower stratosphere as well.

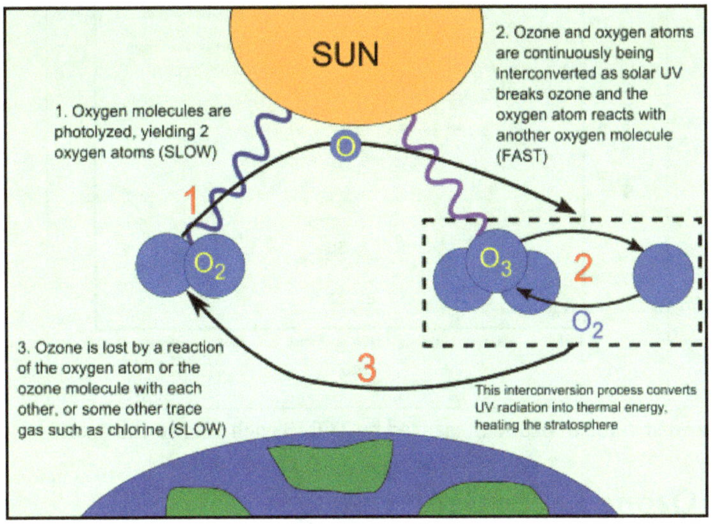

The ozone cycle.

A single chlorine atom would keep on destroying ozone (thus a catalyst) for up to two years (the time scale for transport back down to the troposphere) were it not for reactions that remove them from this cycle by forming reservoir species such as hydrogen chloride (HCl) and chlorine nitrate (ClONO2). On a per atom basis, bromine is even more efficient than chlorine at destroying ozone, but there is much less bromine in the atmosphere at present. As a result, both chlorine and bromine contribute significantly to overall ozone depletion. Laboratory studies have shown that fluorine and iodine atoms participate in analogous catalytic cycles. However, in the Earth's stratosphere, fluorine atoms react rapidly with water and methane to form strongly bound HF, while organic molecules containing iodine react so rapidly in the lower atmosphere that they do not reach the stratosphere in significant quantities.

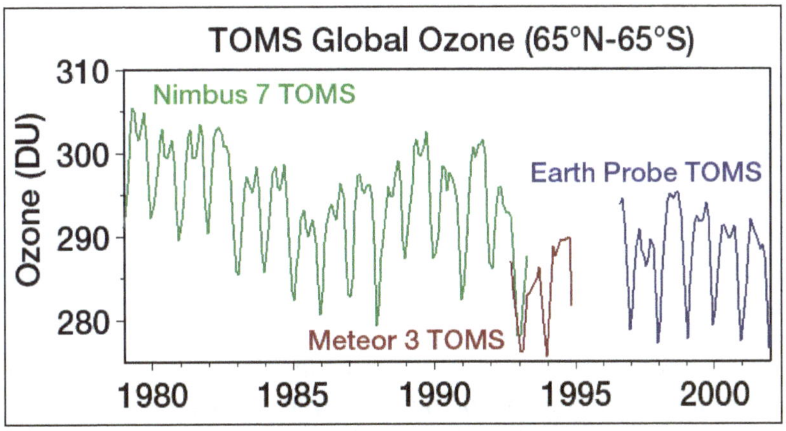

Global monthly average total ozone amount.

On average, a single chlorine atom is able to react with 100,000 ozone molecules before it is removed from the catalytic cycle. This fact plus the amount of chlorine released into the atmosphere yearly by chlorofluorocarbons (CFCs) and hydrochlorofluorocarbons (HCFCs) demonstrates how dangerous CFCs and HCFCs are to the environment.

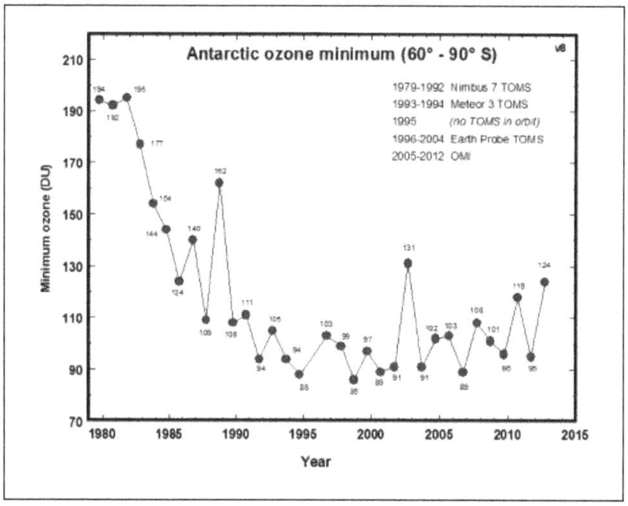

Lowest value of ozone measured by TOMS each year in the ozone hole.

Observations on Ozone Layer Depletion

The most-pronounced decrease in ozone has been in the lower stratosphere. However, the ozone hole is most usually measured not in terms of ozone concentrations at these levels (which are typically a few parts per million) but by reduction in the total *column ozone* above a point on the Earth's surface, which is normally expressed in Dobson units, abbreviated as "DU". Marked decreases in column ozone in the Antarctic spring and early summer compared to the early 1970s and before have been observed using instruments such as the Total Ozone Mapping Spectrometer (TOMS).

Reductions of up to 70 percent in the ozone column observed in the austral (southern hemispheric) spring over Antarctica and first reported in 1985 (Farman et al.) are continuing. Since the 1990s, Antarctic total column ozone in September and October continued to be 40–50 percent lower than pre-ozone-hole values. A gradual trend toward "healing" was reported in 2016.

In the Arctic, the amount lost is more variable year-to-year than in the Antarctic. The greatest Arctic declines, up to 30 percent, are in the winter and spring, when the stratosphere is coldest.

Reactions that take place on polar stratospheric clouds (PSCs) play an important role in enhancing ozone depletion. PSCs form more readily in the extreme cold of the Arctic and Antarctic stratosphere. This is why ozone holes first formed, and are deeper, over Antarctica. Early models failed to take PSCs into account and predicted a gradual global depletion, which is why the sudden Antarctic ozone hole was such a surprise to many scientists.

In middle latitudes, it is more accurate to speak of ozone depletion rather than holes. Total column ozone declined to about six percent below pre-1980 values between 1980 and 1996 for the mid-latitudes of 35–60°N and 35–60°S. In the northern mid-latitudes, it then increased from the minimum value by about two percent from 1996 to 2009 as regulations took effect and the amount of chlorine in the stratosphere decreased. In the Southern Hemisphere's mid-latitudes, total ozone remained constant over that time period. In the tropics, there are no significant trends, largely because halogen-containing compounds have not yet had time to break down and release chlorine and bromine atoms at tropical latitudes.

Large volcanic eruptions have been shown to have substantial albeit uneven ozone-depleting effects, as observed (for example) with the 1991 eruption of Mt. Pinotubo in the Philippines.

Ozone depletion also explains much of the observed reduction in stratospheric and upper tropospheric temperatures. The source of the warmth of the stratosphere is the absorption of UV radiation by ozone, hence reduced ozone leads to cooling. Some stratospheric cooling is also predicted from increases in greenhouse gases such as CO2 and CFCs themselves; however the ozone-induced cooling appears to be dominant.

Predictions of ozone levels remain difficult, but the precision of models' predictions of observed values and the agreement among different modeling techniques have increased steadily. The World Meteorological Organization Global Ozone Research and Monitoring Project—Report No. 44 comes out strongly in favor of the Montreal Protocol, but notes that a UNEP 1994 Assessment overestimated ozone loss for the 1994–1997 period.

Chemicals in the Atmosphere

CFCs and Related Compounds in the Atmosphere

Chlorofluorocarbons (CFCs) and other halogenated ozone depleting substances (ODS) are mainly responsible for man-made chemical ozone depletion. The total amount of effective halogens (chlorine and bromine) in the stratosphere can be calculated and are known as the equivalent effective stratospheric chlorine (EESC).

CFCs were invented by Thomas Midgley, Jr. in the 1920s. They were used in air conditioning and cooling units, as aerosol spray propellants prior to the 1970s, and in the cleaning processes of delicate electronic equipment. They also occur as by-products of some chemical processes. No significant natural sources have ever been identified for these compounds—their presence in the atmosphere is due almost entirely to human manufacture. As mentioned above, when such ozone-depleting chemicals reach the stratosphere, they are dissociated by ultraviolet light to release chlorine atoms. The chlorine atoms act as a catalyst, and each can break down tens of thousands of ozone molecules before being removed from the stratosphere. Given the longevity of CFC molecules, recovery times are measured in decades. It is calculated that a CFC molecule takes an average of about five to seven years to go from the ground level up to the upper atmosphere, and it can stay there for about a century, destroying up to one hundred thousand ozone molecules during that time.

1,1,1-Trichloro-2,2,2-trifluoroethane, also known as CFC-113a, is one of four man-made chemicals newly discovered in the atmosphere by a team at the University of East Anglia. CFC-113a is the only known CFC whose abundance in the atmosphere is still growing. Its source remains a mystery, but illegal manufacturing is suspected by some. CFC-113a seems to have been accumulating unabated since 1960. Between 2010 and 2012, emissions of the gas jumped by 45 percent.

Computer Modeling

Scientists have been increasingly able to attribute the observed ozone depletion to the increase of man-made (anthropogenic) halogen compounds from CFCs by the use of complex chemistry transport models and their validation against observational data (e.g. SLIMCAT, CLaMS—Chemical Lagrangian

Model of the Stratosphere). These models work by combining satellite measurements of chemical concentrations and meteorological fields with chemical reaction rate constants obtained in lab experiments. They are able to identify not only the key chemical reactions but also the transport processes that bring CFC photolysis products into contact with ozone.

Ozone Hole and its Causes

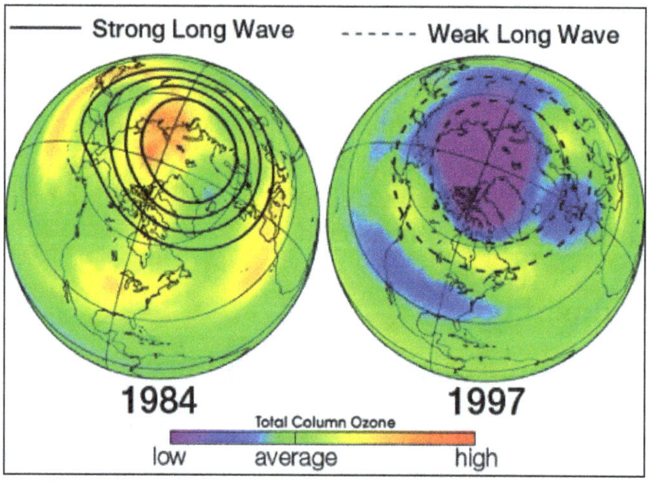

Ozone hole in North America during 1984 (abnormally warm reducing ozone depletion) and 1997 (abnormally cold resulting in increased seasonal depletion).

The Antarctic ozone hole is an area of the Antarctic stratosphere in which the recent ozone levels have dropped to as low as 33 percent of their pre-1975 values. The ozone hole occurs during the Antarctic spring, from September to early December, as strong westerly winds start to circulate around the continent and create an atmospheric container. Within this polar vortex, over 50 percent of the lower stratospheric ozone is destroyed during the Antarctic spring.

As explained above, the primary cause of ozone depletion is the presence of chlorine-containing source gases (primarily CFCs and related halocarbons). In the presence of UV light, these gases dissociate, releasing chlorine atoms, which then go on to catalyze ozone destruction. The Cl-catalyzed ozone depletion can take place in the gas phase, but it is dramatically enhanced in the presence of polar stratospheric clouds (PSCs).

These polar stratospheric clouds (PSC) form during winter, in the extreme cold. Polar winters are dark, consisting of 3 months without solar radiation (sunlight). The lack of sunlight contributes to a decrease in temperature and the polar vortex traps and chills air. Temperatures hover around or below −80 °C. These low temperatures form cloud particles. There are three types of PSC clouds—nitric acid trihydrate clouds, slowly cooling water-ice clouds, and rapid cooling water-ice (nacerous) clouds—provide surfaces for chemical reactions whose products will, in the spring lead to ozone destruction.

The photochemical processes involved are complex but well understood. The key observation is that, ordinarily, most of the chlorine in the stratosphere resides in "reservoir" compounds, primarily chlorine nitrate (ClONO2) as well as stable end products such as HCl. The formation of end products essentially remove Cl from the ozone depletion process. The former sequester Cl, which can be later made available via absorption of light at shorter wavelengths than 400 nm. During the Antarctic winter and spring, however, reactions on the surface of the polar stratospheric cloud particles convert these "reservoir" compounds into reactive free radicals (Cl and ClO). The process by which the clouds remove NO2 from the stratosphere by converting it to nitric acid in the PSC particles, which then are lost by sedimentation is called denitrification. This prevents newly formed ClO from being converted back into ClONO2.

The role of sunlight in ozone depletion is the reason why the Antarctic ozone depletion is greatest during spring. During winter, even though PSCs are at their most abundant, there is no light over the pole to drive chemical reactions. During the spring, however, the sun comes out, providing energy to drive photochemical reactions and melt the polar stratospheric clouds, releasing considerable ClO, which drives the hole mechanism. Further warming temperatures near the end of spring break up the vortex around mid-December. As warm, ozone and NO 2-rich air flows in from lower latitudes, the PSCs are destroyed, the enhanced ozone depletion process shuts down, and the ozone hole closes.

Most of the ozone that is destroyed is in the lower stratosphere, in contrast to the much smaller ozone depletion through homogeneous gas phase reactions, which occurs primarily in the upper stratosphere.

Interest in Ozone Layer Depletion

Public misconceptions and misunderstandings of complex issues like the ozone depletion are common. The limited scientific knowledge of the public led to a confusion with global warming or the perception of global warming as a subset of the "ozone hole". In the beginning, classical green NGOs refrained from using CFC depletion for campaigning, as they assumed the topic was too complicated. They became active much later, e.g. in Greenpeace's support for a CFC-free fridge produced by the former East German company VEB dkk Scharfenstein.

The metaphors used in the CFC discussion (ozone shield, ozone hole) are not "exact" in the scientific sense. The "ozone hole" is more of a *depression*, less "a hole in the windshield". The ozone does not disappear through the layer, nor is there a uniform "thinning" of the ozone layer. However they resonated better with non-scientists and their concerns. The ozone hole was seen as a "hot issue" and imminent risk as lay people feared severe personal consequences such skin cancer, cataracts, damage to plants, and reduction of plankton populations in the ocean's photic zone. Not only on the policy level, ozone regulation compared to climate change fared much better in public opinion. Americans voluntarily switched away from aerosol sprays before legislation was enforced, while climate change failed to achieve comparable concern and public action. The sudden recognition in 1985 that there was a substantial "hole" was widely reported in the press. The especially rapid ozone depletion in Antarctica had previously been dismissed as a measurement error. Scientific consensus was established after regulation.

While the Antarctic ozone hole has a relatively small effect on global ozone, the hole has generated a great deal of public interest because:

- Many have worried that ozone holes might start appearing over other areas of the globe, though to date the only other large-scale depletion is a smaller ozone "dimple" observed during the Arctic spring around the North Pole. Ozone at middle latitudes has declined, but by a much smaller extent (about 4–5 percent decrease).

- If stratospheric conditions become more severe (cooler temperatures, more clouds, more active chlorine), global ozone may decrease at a greater pace. Standard global warming theory predicts that the stratosphere will cool.
- When the Antarctic ozone hole breaks up each year, the ozone-depleted air drifts out into nearby regions. Decreases in the ozone level of up to 10 percent have been reported in New Zealand in the month following the breakup of the Antarctic ozone hole, with ultraviolet-B radiation intensities increasing by more than 15 percent since the 1970s.

Consequences of Ozone Layer Depletion

Since the ozone layer absorbs UVB ultraviolet light from the sun, ozone layer depletion increases surface UVB levels (all else equal), which could lead to damage, including increase in skin cancer. This was the reason for the Montreal Protocol. Although decreases in stratospheric ozone are well-tied to CFCs and to increases in surface UVB, there is no direct observational evidence linking ozone depletion to higher incidence of skin cancer and eye damage in human beings. This is partly because UVA, which has also been implicated in some forms of skin cancer, is not absorbed by ozone, and because it is nearly impossible to control statistics for lifestyle changes over time.

Increased UV

Ozone, while a minority constituent in Earth's atmosphere, is responsible for most of the absorption of UVB radiation. The amount of UVB radiation that penetrates through the ozone layer decreases exponentially with the slant-path thickness and density of the layer. When stratospheric ozone levels decrease, higher levels of UVB reach the Earth's surface. UV-driven phenolic formation in tree rings has dated the start of ozone depletion in northern latitudes to the late 1700s.

In October 2008, the Ecuadorian Space Agency published a report called HIPERION, a study of the last 28 years data from 10 satellites and dozens of ground instruments around the world among them their own, and found that the UV radiation reaching equatorial latitudes was far greater than expected, with the UV Index climbing as high as 24 in some very populated cities; the WHO considers 11 as an extreme index and a great risk to health. The report concluded that depleted ozone levels around the mid-latitudes of the planet are already endangering large populations in these areas. Later, the CONIDA, the Peruvian Space Agency, published its own study, which yielded almost the same findings as the Ecuadorian study.

Biological Effects

The main public concern regarding the ozone hole has been the effects of increased surface UV radiation on human health. So far, ozone depletion in most locations has been typically a few percent and, as noted above, no direct evidence of health damage is available in most latitudes. If the high levels of depletion seen in the ozone hole were to be common across the globe, the effects could be substantially more dramatic. As the ozone hole over Antarctica has in some instances grown so large as to affect parts of Australia, New Zealand, Chile, Argentina, and South Africa, environmentalists have been concerned that the increase in surface UV could be significant.

Ozone depletion would magnify all of the effects of UV on human health, both positive (including production of Vitamin D) and negative (including sunburn, skin cancer, and cataracts). In addition, increased surface UV leads to increased tropospheric ozone, which is a health risk to humans.

Basal and Squamous Cell Carcinomas

The most common forms of skin cancer in humans, basal and squamous cell carcinomas, have been strongly linked to UVB exposure. The mechanism by which UVB induces these cancers is well understood—absorption of UVB radiation causes the pyrimidine bases in the DNA molecule to form dimers, resulting in transcription errors when the DNA replicates. These cancers are relatively mild and rarely fatal, although the treatment of squamous cell carcinoma sometimes requires extensive reconstructive surgery. By combining epidemiological data with results of animal studies, scientists have estimated that every one percent decrease in long-term stratospheric ozone would increase the incidence of these cancers by two percent.

Malignant Melanoma

Another form of skin cancer, malignant melanoma, is much less common but far more dangerous, being lethal in about 15–20 percent of the cases diagnosed. The relationship between malignant melanoma and ultraviolet exposure is not yet fully understood, but it appears that both UVB and UVA are involved. Because of this uncertainty, it is difficult to estimate the effect of ozone depletion on melanoma incidence. One study showed that a 10 percent increase in UVB radiation was associated with a 19 percent increase in melanomas for men and 16 percent for women. A study of people in Punta Arenas, at the southern tip of Chile, showed a 56 percent increase in melanoma and a 46 percent increase in nonmelanoma skin cancer over a period of seven years, along with decreased ozone and increased UVB levels.

Cortical Cataracts

Epidemiological studies suggest an association between ocular cortical cataracts and UVB exposure, using crude approximations of exposure and various cataract assessment techniques. A detailed assessment of ocular exposure to UVB was carried out in a study on Chesapeake Bay Watermen, where increases in average annual ocular exposure were associated with increasing risk of cortical opacity. In this highly exposed group of predominantly white males, the evidence linking cortical opacities to sunlight exposure was the strongest to date. Based on these results, ozone depletion is predicted to cause hundreds of thousands of additional cataracts by 2050.

Increased Tropospheric Ozone

Increased surface UV leads to increased tropospheric ozone. Ground-level ozone is generally recognized to be a health risk, as ozone is toxic due to its strong oxidant properties. The risks are particularly high for young children, the elderly, and those with asthma or other respiratory difficulties. At this time, ozone at ground level is produced mainly by the action of UV radiation on combustion gases from vehicle exhausts.

Increased Production of Vitamin D

Vitamin D is produced in the skin by ultraviolet light. Thus, higher UVB exposure raises human vitamin D in those deficient in it. Recent research (primarily since the Montreal Protocol) shows that many humans have less than optimal vitamin D levels. In particular, in the U.S. population, the lowest quarter of vitamin D (<17.8 ng/ml) were found using information from the National Health and Nutrition Examination Survey to be associated with an increase in all-cause mortality in the general population. While blood level of Vitamin D in excess of 100 ng/ml appear to raise blood calcium excessively and to be associated with higher mortality, the body has mechanisms that prevent sunlight from producing Vitamin D in excess of the body's requirements.

Effects on Non-human Animals

A November 2010 report by scientists at the Institute of Zoology in London found that whales off the coast of California have shown a sharp rise in sun damage, and these scientists "fear that the thinning ozone layer is to blame". The study photographed and took skin biopsies from over 150 whales in the Gulf of California and found "widespread evidence of epidermal damage commonly associated with acute and severe sunburn", having cells that form when the DNA is damaged by UV radiation. The findings suggest "rising UV levels as a result of ozone depletion are to blame for the observed skin damage, in the same way that human skin cancer rates have been on the increase in recent decades."

Effects on Crops

An increase of UV radiation would be expected to affect crops. A number of economically important species of plants, such as rice, depend on cyanobacteria residing on their roots for the retention of nitrogen. Cyanobacteria are sensitive to UV radiation and would be affected by its increase. "Despite mechanisms to reduce or repair the effects of increased ultraviolet radiation, plants have a limited ability to adapt to increased levels of UVB, therefore plant growth can be directly affected by UVB radiation."

Public Policy

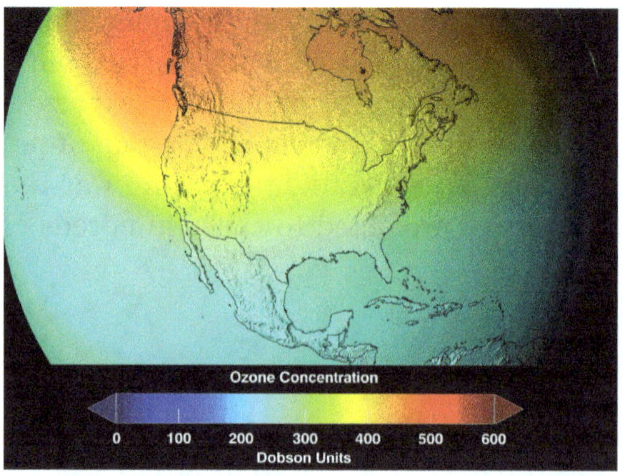

NASA projections of stratospheric ozone concentrations if chlorofluorocarbons had not been banned.

The full extent of the damage that CFCs have caused to the ozone layer is not known and will not be known for decades; however, marked decreases in column ozone have already been observed. The

Montreal and Vienna conventions were installed long before a scientific consensus was established or important uncertainties in the science field were being resolved. The ozone case was understood comparably well by lay persons as e.g. *Ozone shield* or *ozone hole* were useful "easy-to-understand bridging metaphors". Americans voluntarily switched away from aerosol sprays, resulting in a 50 percent sales loss even before legislation was enforced.

After a 1976 report by the United States National Academy of Sciences concluded that credible scientific evidence supported the ozone depletion hypothesis a few countries, including the United States, Canada, Sweden, Denmark, and Norway, moved to eliminate the use of CFCs in aerosol spray cans. At the time this was widely regarded as a first step towards a more comprehensive regulation policy, but progress in this direction slowed in subsequent years, due to a combination of political factors (continued resistance from the halocarbon industry and a general change in attitude towards environmental regulation during the first two years of the Reagan administration) and scientific developments (subsequent National Academy assessments that indicated that the first estimates of the magnitude of ozone depletion had been overly large). A critical DuPont manufacturing patent for Freon was set to expire in 1979. The United States banned the use of CFCs in aerosol cans in 1978. The European Community rejected proposals to ban CFCs in aerosol sprays, and in the U.S., CFCs continued to be used as refrigerants and for cleaning circuit boards. Worldwide CFC production fell sharply after the U.S. aerosol ban, but by 1986 had returned nearly to its 1976 level. In 1993, DuPont shut down its CFC facility.

The U.S. Government's attitude began to change again in 1983, when William Ruckelshaus replaced Anne M. Burford as Administrator of the United States Environmental Protection Agency. Under Ruckelshaus and his successor, Lee Thomas, the EPA pushed for an international approach to halocarbon regulations. In 1985 20 nations, including most of the major CFC producers, signed the Vienna Convention for the Protection of the Ozone Layer, which established a framework for negotiating international regulations on ozone-depleting substances. That same year, the discovery of the Antarctic ozone hole was announced, causing a revival in public attention to the issue. In 1987, representatives from 43 nations signed the Montreal Protocol. Meanwhile, the halocarbon industry shifted its position and started supporting a protocol to limit CFC production. However, this shift was uneven with DuPont acting more quickly than their European counterparts. DuPont may have feared court action related to increased skin cancer especially as the EPA had published a study in 1986 claiming that an additional 40 million cases and 800,000 cancer deaths were to be expected in the U.S. in the next 88 years. The EU shifted its position as well after Germany gave up its defence of the CFC industry and started supporting moves towards regulation. Government and industry in France and the UK tried to defend their CFC producing industries even after the Montreal Protocol had been signed.

At Montreal, the participants agreed to freeze production of CFCs at 1986 levels and to reduce production by 50 percent by 1999. After a series of scientific expeditions to the Antarctic produced convincing evidence that the ozone hole was indeed caused by chlorine and bromine from manmade organohalogens, the Montreal Protocol was strengthened at a 1990 meeting in London. The participants agreed to phase out CFCs and halons entirely (aside from a very small amount marked for certain "essential" uses, such as asthma inhalers) by 2000 in non-Article 5 countries and by 2010 in Article 5 (less developed) signatories. At a 1992 meeting in Copenhagen, the phase-out date was moved up to 1996. At the same meeting, methyl bromide (MeBr), a fumigant used primarily in agricultural production, was added to the list of controlled substances. For all substances controlled under the protocol, phaseout schedules were delayed for less developed ('Article 5(1)') countries, and phaseout in these countries was supported by transfers of expertise, technology, and money from non-Article 5(1) Parties to the Protocol. Additionally, exemptions from the agreed schedules could be applied for under the Essential Use Exemption (EUE) process for substances other than methyl bromide and under the Critical Use Exemption (CUE) process for methyl bromide.

A hydrocarbon refrigerant was developed in 1992 at the request of the non-governmental organization (NGO) Greenpeace, and was being used by some 40 percent of the refrigerator market in 2013. In the U.S., however, change has been much slower. To some extent, CFCs were being replaced by the less damaging hydrochlorofluorocarbons (HCFCs), although concerns remain regarding HCFCs also. In some applications, hydrofluorocarbons (HFCs) were being used to replace CFCs. HFCs, which contain no chlorine or bromine, do not contribute at all to ozone depletion although they are potent greenhouse gases. The best known of these compounds is probably HFC-134a (R-134a), which in the United States has largely replaced CFC-12 (R-12) in automobile air conditioners. In laboratory analytics (a former "essential" use) the ozone depleting substances can be replaced with various other solvents. The development and promotion of an ozone-safe hydrocarbon refrigerant was a breakthrough which occurred after the initiative of a non-governmental organization (NGO). Civil society including especially NGOs, in fact, played critical roles at all stages of policy development leading up to the Vienna Conference, the Montreal Protocol, and in assessing compliance afterwards. The major companies claimed that no alternatives to HFC existed. By 1992, Greenpeace had requested a scientific team to research an ozone-safe refrigerant, an effort which resulted in a successful hydrocarbon formula, for which Greenpeace left the patent as open source. The NGO maintained the technology as an open patent as it worked successfully first with a small company to market an appliance beginning in Europe, then Asia and later Latin America, receiving a 1997 UNEP award. By 1995, Germany had already made CFC refrigerators illegal. Production spread to companies like Electrolux, Bosch, and LG, with sales reaching some 300 million refrigerators by 2008. However, the giant corporations all continued to refuse to apply the technology in Latin America. In 2003, a domestic Argentinian company began Greenfreeze production, while the giant Bosch in Brazil began a year later. Chemical companies like Du Pont, whose representatives even disparaged the green technology as "that German technology," maneuvered the EPA to block the technology in the U.S. until 2011. Ben & Jerry's of Unilever and General Electric, spurred by Greenpeace, had expressed formal interest in 2008 which figured in the EPA's final approval. Currently, more than 600 million refrigerators have been sold, making up more than 40 percent of the market. Since 2004, corporations like Coca-Cola, Carlsberg, and IKEA have been forming a coalition to promote the ozone-safe Greenfreeze units.

More recently, policy experts have advocated for efforts to link ozone protection efforts to climate protection efforts. Many ODS are also greenhouse gases, some thousands of times more powerful agents of radiative forcing than carbon dioxide over the short and medium term. Thus policies protecting the ozone layer have had benefits in mitigating climate change. In fact, the reduction of the radiative forcing due to ODS probably masked the true level of climate change effects of other GHGs, and was responsible for the "slow down" of global warming from the mid-90s. Policy decisions in one arena affect the costs and effectiveness of environmental improvements in the other.

ODS Requirements in the Marine Industry

The IMO has amended MARPOL Annex VI Regulation 12 regarding ozone depleting substances. As from July 1, 2010, all vessels where MARPOL Annex VI is applicable should have a list of equipment using ozone depleting substances. The list should include name of ODS, type and location of equipment, quantity in kg and date. All changes since that date should be recorded in an ODS Record book on board recording all intended or unintended releases to the atmosphere. Furthermore, new ODS supply or landing to shore facilities should be recorded as well.

Prospects of Ozone Depletion

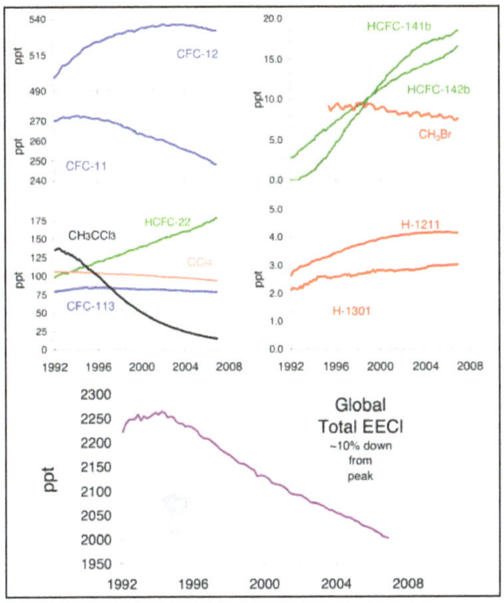

Ozone-depleting gas trends.

Since the adoption and strengthening of the Montreal Protocol has led to reductions in the emissions of CFCs, atmospheric concentrations of the most-significant compounds have been declining. These substances are being gradually removed from the atmosphere; since peaking in 1994, the Effective Equivalent Chlorine (EECl) level in the atmosphere had dropped about 10 percent by 2008. The decrease in ozone-depleting chemicals has also been significantly affected by a decrease in bromine-containing chemicals. The data suggest that substantial natural sources exist for atmospheric methyl bromide (CH3Br). The phase-out of CFCs means that nitrous oxide (N2O), which is not covered by the Montreal Protocol, has become the most highly emitted ozone-depleting substance and is expected to remain so throughout the 21st century.

A 2005 IPCC review of ozone observations and model calculations concluded that the global amount of ozone has now approximately stabilized. Although considerable variability is expected from year to year, including in polar regions where depletion is largest, the ozone layer is expected to begin to recover in coming decades due to declining ozone-depleting substance concentrations, assuming full compliance with the Montreal Protocol.

The Antarctic ozone hole is expected to continue for decades. Ozone concentrations in the lower stratosphere over Antarctica will increase by 5–10 percent by 2020 and return to pre-1980 levels by about 2060–2075. This is 10–25 years later than predicted in earlier assessments, because of revised estimates of atmospheric concentrations of ozone-depleting substances, including a larger predicted future usage in developing countries. Another factor that may prolong ozone depletion is the drawdown of nitrogen oxides from above the stratosphere due to changing wind patterns. A gradual trend toward "healing" was reported in 2016.

Research History

The basic physical and chemical processes that lead to the formation of an ozone layer in the Earth's stratosphere were discovered by Sydney Chapman in 1930. Short-wavelength UV radiation splits an oxygen (O_2) molecule into two oxygen (O) atoms, which then combine with other oxygen molecules to form ozone. Ozone is removed when an oxygen atom and an ozone molecule "recombine" to form two oxygen molecules, i.e. $O + O_3 \rightarrow 2O_2$. In the 1950s, David Bates and Marcel Nicolet presented evidence that various free radicals, in particular hydroxyl (OH) and nitric oxide (NO), could catalyze this recombination reaction, reducing the overall amount of ozone. These free radicals were known to be present in the stratosphere, and so were regarded as part of the natural balance–it was estimated that in their absence, the ozone layer would be about twice as thick as it currently is.

In 1970 Paul Crutzen pointed out that emissions of nitrous oxide (N_2O), a stable, long-lived gas produced by soil bacteria, from the Earth's surface could affect the amount of nitric oxide (NO) in the stratosphere. Crutzen showed that nitrous oxide lives long enough to reach the stratosphere, where it is converted into NO. Crutzen then noted that increasing use of fertilizers might have led to an increase in nitrous oxide emissions over the natural background, which would in turn result in an increase in the amount of NO in the stratosphere. Thus human activity could affect the stratospheric ozone layer. In the following year, Crutzen and (independently) Harold Johnston suggested that NO emissions from supersonic passenger aircraft, which would fly in the lower stratosphere, could also deplete the ozone layer. However, more recent analysis in 1995 by David W. Fahey, an atmospheric scientist at the National Oceanic and Atmospheric Administration, found that the drop in ozone would be from 1–2 percent if a fleet of 500 supersonic passenger aircraft were operated. This, Fahey expressed, would not be a showstopper for advanced supersonic passenger aircraft development.

Rowland–molina Hypothesis

In 1974 Frank Sherwood Rowland, Chemistry Professor at the University of California at Irvine, and his postdoctoral associate Mario J. Molina suggested that long-lived organic halogen compounds, such as CFCs, might behave in a similar fashion as Crutzen had proposed for nitrous oxide. James Lovelock had recently discovered, during a cruise in the South Atlantic in 1971, that almost all of the CFC compounds manufactured since their invention in 1930 were still present in the atmosphere. Molina and Rowland concluded that, like N_2O, the CFCs would reach the stratosphere where they would be dissociated by UV light, releasing chlorine atoms. A year earlier, Richard Stolarski and Ralph Cicerone at the University of Michigan had shown that Cl is even more efficient than NO at catalyzing the destruction of ozone. Similar conclusions were reached by Michael McElroy and Steven Wofsy at Harvard University. Neither group, however, had realized that CFCs were a potentially large source of stratospheric chlorine–instead, they had been investigating the possible effects of HCl emissions from the Space Shuttle, which are very much smaller.

The Rowland–Molina hypothesis was strongly disputed by representatives of the aerosol and halocarbon industries. The Chair of the Board of DuPont was quoted as saying that ozone depletion theory is "a science fiction tale ... a load of rubbish ... utter nonsense". Robert Abplanalp, the President of Precision Valve Corporation (and inventor of the first practical aerosol spray can valve), wrote to the Chancellor of UC Irvine to complain about Rowland's public statements. Nevertheless, within three years most of the basic assumptions made by Rowland and Molina were confirmed by laboratory measurements and by direct observation in the stratosphere. The concentrations of the source gases (CFCs and related compounds) and the chlorine reservoir species (HCl and ClONO2) were measured throughout the stratosphere, and demonstrated that CFCs were indeed the major source of stratospheric chlorine, and that nearly all of the CFCs emitted would eventually reach the stratosphere. Even more convincing was the measurement, by James G. Anderson and collaborators, of chlorine monoxide (ClO) in the stratosphere. ClO is produced by the reaction of Cl with ozone—its observation thus demonstrated that Cl radicals not only were present in the stratosphere but also were actually involved in destroying ozone. McElroy and Wofsy extended the work of Rowland and Molina by showing that bromine atoms were even more effective catalysts for ozone loss than chlorine atoms and argued that the brominated organic compounds known as halons, widely used in fire extinguishers, were a potentially large source of stratospheric bromine. In 1976 the United States National Academy of Sciences released a report concluding that the ozone depletion hypothesis was strongly supported by the scientific evidence. Scientists calculated that if CFC production continued to increase at the going rate of 10 percent per year until 1990 and then remain steady, CFCs would cause a global ozone loss of 5–7 percent by 1995, and a 30–50 percent loss by 2050. In response the United States, Canada and Norway banned the use of CFCs in aerosol spray cans in 1978. However, subsequent research, summarized by the National Academy in reports issued between 1979 and 1984, appeared to show that the earlier estimates of global ozone loss had been too large.

Crutzen, Molina, and Rowland were awarded the 1995 Nobel Prize in Chemistry for their work on stratospheric ozone.

Antarctic Ozone Hole

The discovery of the Antarctic "ozone hole" by British Antarctic Survey scientists Farman, Gardiner and Shanklin (first reported in a paper in *Nature* in May 1985) came as a shock to the scientific community, because the observed decline in polar ozone was far larger than anyone had anticipated. Satellite measurements showing massive depletion of ozone around the south pole were becoming available at the same time. However, these were initially rejected as unreasonable by data quality control algorithms (they were filtered out as errors since the values were unexpectedly low); the ozone hole was detected only in satellite data when the raw data was reprocessed following evidence of ozone depletion in *in situ* observations. When the software was rerun without the flags, the ozone hole was seen as far back as 1976.

Susan Solomon, an atmospheric chemist at the National Oceanic and Atmospheric Administration (NOAA), proposed that chemical reactions on polar stratospheric clouds (PSCs) in the cold Antarctic stratosphere caused a massive, though localized and seasonal, increase in the amount of chlorine present in active, ozone-destroying forms. The polar stratospheric clouds in Antarctica are only formed when there are very low temperatures, as low as −80 °C, and early spring conditions.

In such conditions the ice crystals of the cloud provide a suitable surface for conversion of unreactive chlorine compounds into reactive chlorine compounds, which can deplete ozone easily.

Moreover, the polar vortex formed over Antarctica is very tight and the reaction occurring on the surface of the cloud crystals is far different from when it occurs in atmosphere. These conditions have led to ozone hole formation in Antarctica. This hypothesis was decisively confirmed, first by laboratory measurements and subsequently by direct measurements, from the ground and from high-altitude airplanes, of very high concentrations of chlorine monoxide (ClO) in the Antarctic stratosphere.

Alternative hypotheses, which had attributed the ozone hole to variations in solar UV radiation or to changes in atmospheric circulation patterns, were also tested and shown to be untenable.

Meanwhile, analysis of ozone measurements from the worldwide network of ground-based Dobson spectrophotometers led an international panel to conclude that the ozone layer was in fact being depleted, at all latitudes outside of the tropics. These trends were confirmed by satellite measurements. As a consequence, the major halocarbon-producing nations agreed to phase out production of CFCs, halons, and related compounds, a process that was completed in 1996.

Since 1981 the United Nations Environment Programme, under the auspices of the World Meteorological Organization, has sponsored a series of technical reports on the Scientific Assessment of Ozone Depletion, based on satellite measurements. The 2007 report showed that the hole in the ozone layer was recovering and the smallest it had been for about a decade. The 2010 report found, "Over the past decade, global ozone and ozone in the Arctic and Antarctic regions is no longer decreasing but is not yet increasing. The ozone layer outside the Polar regions is projected to recover to its pre-1980 levels some time before the middle of this century. In contrast, the springtime ozone hole over the Antarctic is expected to recover much later." In 2012, NOAA and NASA reported "Warmer air temperatures high above the Antarctic led to the second smallest season ozone hole in 20 years averaging 17.9 million square kilometres. The hole reached its maximum size for the season on Sept 22, stretching to 21.2 million square kilometres." A gradual trend toward "healing" was reported in 2016.

The hole in the Earth's ozone layer over the South Pole has affected atmospheric circulation in the Southern Hemisphere all the way to the equator. The ozone hole has influenced atmospheric circulation all the way to the tropics and increased rainfall at low, subtropical latitudes in the Southern Hemisphere.

Arctic Ozone Hole

On March 15, 2011, a record ozone layer loss was observed, with about half of the ozone present over the Arctic having been destroyed. The change was attributed to increasingly cold winters in the Arctic stratosphere at an altitude of approximately 20 km (12 mi), a change associated with global warming in a relationship that is still under investigation. By March 25, the ozone loss had become the largest compared to that observed in all previous winters with the possibility that it would become an ozone hole. This would require that the quantities of ozone to fall below 200 Dobson units, from the 250 recorded over central Siberia. It is predicted that the thinning layer would affect parts of Scandinavia and Eastern Europe on March 30–31.

On March 3, 2005, the journal Nature published an article linking 2004's unusually large Arctic ozone hole to solar wind activity.

On October 2, 2011, a study was published in the journal *Nature*, which said that between December 2010 and March 2011 up to 80 percent of the ozone in the atmosphere at about 20 kilometres (12 mi) above the surface was destroyed. The level of ozone depletion was severe enough that scientists said it could be compared to the ozone hole that forms over Antarctica every winter. According to the study, "for the first time, sufficient loss occurred to reasonably be described as an Arctic ozone hole." The study analyzed data from the Aura and CALIPSO satellites, and determined that the larger-than-normal ozone loss was due to an unusually long period of cold weather in the Arctic, some 30 days more than typical, which allowed for more ozone-destroying chlorine compounds to be created. According to Lamont Poole, a co-author of the study, cloud and aerosol particles on which the chlorine compounds are found "were abundant in the Arctic until mid March 2011—much later than usual—with average amounts at some altitudes similar to those observed in the Antarctic, and dramatically larger than the near-zero values seen in March in most Arctic winters".

Tibet Ozone Hole

As winters that are colder are more affected, at times there is an ozone hole over Tibet. In 2006, a 2.5 million square kilometer ozone hole was detected over Tibet. Also again in 2011 an ozone hole appeared over mountainous regions of Tibet, Xinjiang, Qinghai and the Hindu Kush, along with an unprecedented hole over the Arctic, though the Tibet one is far less intense than the ones over the Arctic or Antarctic.

Potential Depletion by Storm Clouds

Research in 2012 showed that the same process that produces the ozone hole over Antarctica occurs over summer storm clouds in the United States, and thus may be destroying ozone there as well.

Ozone Depletion and Global Warming

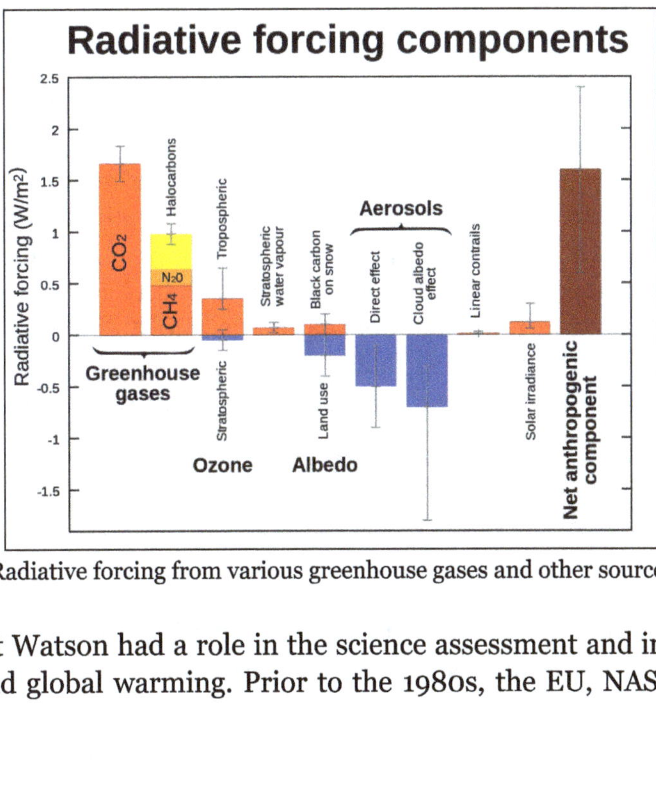

Radiative forcing from various greenhouse gases and other sources.

Among others, Robert Watson had a role in the science assessment and in the regulation efforts of ozone depletion and global warming. Prior to the 1980s, the EU, NASA, NAS, UNEP, WMO and the British government had dissenting scientific reports and Watson played a crucial role in the process of unified assessments. Based on the experience with the ozone case, the IPCC started to work on a unified reporting and science assessment to reach a consensus to provide the IPCC Summary for Policymakers.

There are various areas of linkage between ozone depletion and global warming science:

- The same CO_2 radiative forcing that produces global warming is expected to cool the stratosphere. This cooling, in turn, is expected to produce a relative *increase* in ozone (O_3) depletion in polar area and the frequency of ozone holes.
- Conversely, ozone depletion represents a radiative forcing of the climate system. There are two opposing effects: Reduced ozone causes the stratosphere to absorb less solar radiation, thus cooling the stratosphere while warming the troposphere; the resulting colder stratosphere emits less long-wave radiation downward, thus cooling the troposphere. Overall, the cooling dominates; the IPCC concludes "*observed stratospheric O_3 losses over the past two decades have caused a negative forcing of the surface-troposphere system*" of about -0.15 ± 0.10 watts per square meter (W/m^2).
- One of the strongest predictions of the greenhouse effect is that the stratosphere will cool. Although this cooling has been observed, it is not trivial to separate the effects of changes in the concentration of greenhouse gases and ozone depletion since both will lead to cooling. However, this can be done by numerical stratospheric modeling. Results from the National Oceanic and Atmospheric Administration's Geophysical Fluid Dynamics Laboratory show that above 20 km (12 mi), the greenhouse gases dominate the cooling.
- As noted under 'Public Policy', ozone depleting chemicals are also often greenhouse gases. The increases in concentrations of these chemicals have produced 0.34 ± 0.03 W/m^2 of radiative forcing, corresponding to about 14 percent of the total radiative forcing from increases in the concentrations of well-mixed greenhouse gases.
- The long term modeling of the process, its measurement, study, design of theories and testing take decades to document, gain wide acceptance, and ultimately become the dominant paradigm. Several theories about the destruction of ozone were hypothesized in the 1980s, published in the late 1990s, and are currently being investigated. Dr Drew Schindell, and Dr Paul Newman, NASA Goddard, proposed a theory in the late 1990s, using computational modeling methods to model ozone destruction, that accounted for 78 percent of the ozone destroyed. Further refinement of that model accounted for 89 percent of the ozone destroyed, but pushed back the estimated recovery of the ozone hole from 75 years to 150 years. (An important part of that model is the lack of stratospheric flight due to depletion of fossil fuels.)

Misconceptions

CFC Weight

Since CFC molecules are heavier than air (nitrogen or oxygen), it is commonly believed that the CFC molecules cannot reach the stratosphere in significant amount. However, atmospheric gases are not sorted by weight; the forces of wind can fully mix the gases in the atmosphere. Lighter CFCs are evenly distributed throughout the turbosphere and reach the upper atmosphere, although some of the heavier CFCs are not evenly distributed.

Percentage of Man-made Chlorine

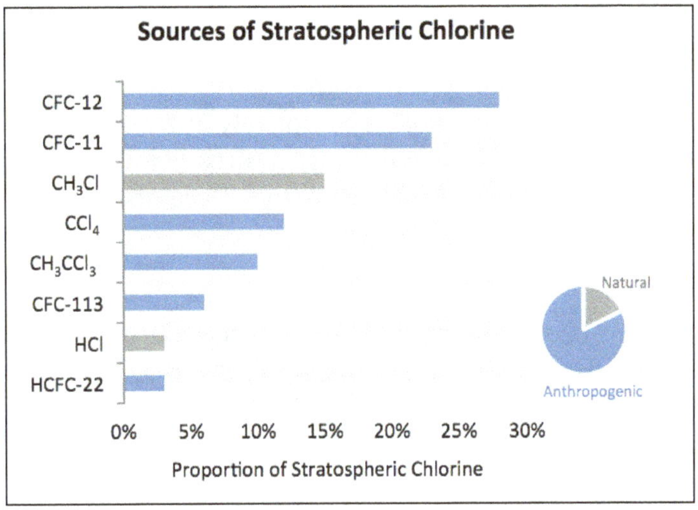

Sources of Stratospheric Chlorine.

Another misconception is that "it is generally accepted that natural sources of tropospheric chlorine are four to five times larger than man-made ones." While strictly true, *tropospheric* chlorine is irrelevant; it is *stratospheric* chlorine that affects ozone depletion. Chlorine from ocean spray is soluble and thus is washed by rainfall before it reaches the stratosphere. CFCs, in contrast, are insoluble and long-lived, allowing them to reach the stratosphere. In the lower atmosphere, there is much more chlorine from CFCs and related haloalkanes than there is in HCl from salt spray, and in the stratosphere halocarbons are dominant. Only methyl chloride, which is one of these halocarbons, has a mainly natural source, and it is responsible for about 20 percent of the chlorine in the stratosphere; the remaining 80 percent comes from manmade sources.

Very violent volcanic eruptions can inject HCl into the stratosphere, but researchers have shown that the contribution is not significant compared to that from CFCs. A similar erroneous assertion is that soluble halogen compounds from the volcanic plume of Mount Erebus on Ross Island, Antarctica are a major contributor to the Antarctic ozone hole.

Nevertheless, the latest study showed that the role of Mount Erebus volcano in the Antarctic ozone depletion was probably underestimated. Based on the NCEP/NCAR reanalysis data over the last 35 years and by using the NOAA HYSPLIT trajectory model, researchers showed that Erebus volcano gas emissions (including hydrogen chloride (HCl)) can reach the Antarctic stratosphere via high-latitude cyclones and then the polar vortex. Depending on Erebus volcano activity, the additional annual HCl mass entering the stratosphere from Erebus varies from 1.0 to 14.3 kt.

First Observation

G.M.B. Dobson (Exploring the Atmosphere, 2nd Edition, Oxford, 1968) mentioned that when springtime ozone levels in the Antarctic over Halley Bay were first measured in 1956, he was surprised to find that they were ~320 DU, or about 150 DU below spring Arctic levels of ~450 DU. These were at that time the only known Antarctic ozone values available. What Dobson describes is essentially the *baseline* from which the ozone hole is measured: actual ozone hole values are in the 150–100 DU range.

The discrepancy between the Arctic and Antarctic noted by Dobson was primarily a matter of timing: during the Arctic spring ozone levels rose smoothly, peaking in April, whereas in the Antarctic they stayed approximately constant during early spring, rising abruptly in November when the polar vortex broke down.

The behavior seen in the Antarctic ozone hole is completely different. Instead of staying constant, early springtime ozone levels suddenly drop from their already low winter values, by as much as 50 percent, and normal values are not reached again until December.

Location of Hole

Some people thought that the ozone hole should be above the sources of CFCs. However, CFCs are well mixed globally in the troposphere and stratosphere. The reason for occurrence of the ozone hole above Antarctica is not because there are more CFCs concentrated but because the low temperatures help form polar stratospheric clouds. In fact, there are findings of significant and localized "ozone holes" above other parts of the earth.

World Ozone Day

In 1994, the United Nations General Assembly voted to designate September 16 as "World Ozone Day", to commemorate the signing of the Montreal Protocol on that date in 1987.

References

- "Reconstruction of Paleobehavior of Ozonosphere Based on Response to UV-B Radiation Effect in Dendrochronologic Signal" (PDF). Atmospheric Radiation Measurement, USA. Retrieved 09, April 2020
- "Climate Change 2001: Working Group I: The Scientific Basis". Intergovernmental Panel on Climate Change Work Group I. 2001. pp. Chapter 6.4 Stratospheric Ozone. Retrieved 15, June 2020
- Fountain, Henry (July 27, 2012). "Storms Threaten Ozone Layer Over U.S., Study Says". The New York Times. p. A1. Retrieved 20, May 2020
- "Lanzan la primera de las "Propuestas Greenpeace": la heladera "Greenfreeze" | Greenpeace Argentina". Greenpeace.org. Retrieved 19, July 2020
- "The Relative Roles of Ozone and Other Greenhouse Gases in Climate Change in the Stratosphere". Geophysical Fluid Dynamics Laboratory. February 29, 2004. Archived from the original on January 20, 2009. Retrieved March 13, 2015. CS1 maint: Unfit url (link)
- "NOAA Study Shows Nitrous Oxide Now Top Ozone-Depleting Emission". Noaanews.noaa.gov. August 27, 2009. Retrieved 14, January 2020
- "Arctic ozone loss at record level". BBC News Online. October 2, 2011. Archived from the original on October 2, 2011. Retrieved 05, April 2020
- "The Arctic Ozone Sieve: More Global Weirding?". Scienceblogs.com. March 25, 2011. Archived from the original on April 4, 2011. Retrieved 17, August 2020

- "Earth news: Chinese Scientists Find New Ozone Hole Over Tibet". Elainemeinelsupkis.typepad.com. May 4, 2006. Retrieved 12, June 2020
- "Nash, Eric; Newman, Paul (September 19, 2001). "NASA Confirms Arctic Ozone Depletion Trigger". Image of the Day. NASA. Retrieved 02, March 2020
- "Twenty Questions and Answers About the Ozone Layer". Scientific Assessment of Ozone Depletion: 2010 (PDF). World Meteorological Organization. 2011. Retrieved 25, August 2020
- Schiermeier, Quirin (February 22, 1999). "The Great Beyond: Arctic ozone hole causes concern". Blogs.nature.com. Retrieved 29, May 2020

Permissions

All chapters in this book are published with permission under the Creative Commons Attribution Share Alike License or equivalent. Every chapter published in this book has been scrutinized by our experts. Their significance has been extensively debated. The topics covered herein carry significant information for a comprehensive understanding. They may even be implemented as practical applications or may be referred to as a beginning point for further studies.

We would like to thank the editorial team for lending their expertise to make the book truly unique. They have played a crucial role in the development of this book. Without their invaluable contributions this book wouldn't have been possible. They have made vital efforts to compile up to date information on the varied aspects of this subject to make this book a valuable addition to the collection of many professionals and students.

This book was conceptualized with the vision of imparting up-to-date and integrated information in this field. To ensure the same, a matchless editorial board was set up. Every individual on the board went through rigorous rounds of assessment to prove their worth. After which they invested a large part of their time researching and compiling the most relevant data for our readers.

The editorial board has been involved in producing this book since its inception. They have spent rigorous hours researching and exploring the diverse topics which have resulted in the successful publishing of this book. They have passed on their knowledge of decades through this book. To expedite this challenging task, the publisher supported the team at every step. A small team of assistant editors was also appointed to further simplify the editing procedure and attain best results for the readers.

Apart from the editorial board, the designing team has also invested a significant amount of their time in understanding the subject and creating the most relevant covers. They scrutinized every image to scout for the most suitable representation of the subject and create an appropriate cover for the book.

The publishing team has been an ardent support to the editorial, designing and production team. Their endless efforts to recruit the best for this project, has resulted in the accomplishment of this book. They are a veteran in the field of academics and their pool of knowledge is as vast as their experience in printing. Their expertise and guidance has proved useful at every step. Their uncompromising quality standards have made this book an exceptional effort. Their encouragement from time to time has been an inspiration for everyone.

The publisher and the editorial board hope that this book will prove to be a valuable piece of knowledge for students, practitioners and scholars across the globe.

Index

www.ingramcontent.com/pod-product-compliance
Lightning Source LLC
LaVergne TN
LVHW080910210926
848485LV00004B/627